Public IT Investment

The Success of IT Projects

Written By

Dr. Abdullah Al-Hatmi

ISBN
978-1-4828-9565-0 (sc)
978-1-4828-9243-7 (e)

To order additional copies of this book, contact
Toll Free 800 101 2657 (Singapore)
Toll Free 1 800 81 7340 (Malaysia)
orders.singapore@partridgepublishing.com

www.partridgepublishing.com/singapore

04/03/2014

PARTRIDGE
A Penguin Random House Company

To my sweetie Mum and in the memory of my Dad, Ali

TABLE OF CONTENTS

5. Analysis of SA Maturity Within Projects

6. Analysis of SA Attributes and Maturity Levels Across Projects

7. Analysis of SA perspectives Maturity Levels Across Projects

8. DISCUSSION AND CONCLUSION

9. REFERENCES

APPENDICES

FIGURES

TABLES

PREFACE

This book pertains to the discussion of Strategic Alignment which, if not implemented properly, can cause the failure of the information technology (IT) projects in an organisation. While some public sector organisations are able to deliver their IT projects on time and on budget, few of them report on project performance and how they actually identify and measure the benefits supposedly arising from those projects (Gershon 2009). IT projects typically involve organisations committing substantial resources and funds, and so project failure can result in not just loss of the funds invested, but also of the benefits that would have accrued directly had the project implementation been better informed.

IT governance strategies and objectives shape Strategic Alignment (SA) perspectives, which in turn affect the management of IT projects. Hence, SA is a key subject in this book. It should be noted that the current trend is to align IT with business as equal partners, not to align IT to fit into a predetermined business strategy. The study identifies strategic ICT alignment issues that contribute to the success of IT projects. Using a local government in Australia as a case study, the book investigates 14 IT projects and highlights how strategic alignment perspectives generate public value through IT projects. The research undertaken in this study has shown that the delivery of IT projects on time and on budget is an inadequate measurement for value realisation from IT. The outcome must also be assessed through alternative measures and governance feedback mechanisms.

The crucial issue that this research investigates is the role of Strategic Alignment (SA) in the success of IT projects in public organisations. It is essential to consider the IT SA perspectives that can be utilised to increase the understanding needed for a better achievement of value realisation in a public organisation. This book develops a conceptual framework as a model of assessment and provides recommendations for its application to government IT projects decision-makers, IT projects and programme managers, and small and medium IT entrepreneurs

Plan of the Book

The book is structured as follows.

Chapter One begins with an overview of the book. It explains the motivation for writing this book and explanation on why IT projects still fail.

Chapter Two contains a literature review on the background of IT, business and governance theory. It discusses strategic alignment perspectives and models and, by identifying gaps in the research, isolates areas that require further analysis in the context of public organisations. The chapter concludes with a suggested conceptual framework to be adapted to a government context.

Chapter Three describes the research design and case study methodology used to investigate the impact of SA perspectives on government IT projects in local governments. The justification for adopting a case study methodology is provided. Sample data collection and analysis procedures are given in detail in this chapter, providing the rationale concerning the validity and reliability of the methods used followed by the operationalisation of the theoretical constructs.

Chapter Four describes the organisational setting of the case study used in this research; a local government. This includes an overview of the organisation, the participants' characteristics, case summary characteristics and the characteristics of the IT projects ongoing in the local government. It also shows how the data was analysed and presented.

Chapter Five introduces the concept of SA maturity levels and discusses this concept in relation to the maturity levels of IT projects. This chapter also examines the success rate of IT projects and the relationship of success to the maturity level of IT projects. Thus, this chapter covers the main analysis, discussion and contribution of this research. Analytical analysis was undertaken further using the NVivo package tool, which helped to manage and analyse data within and across projects.

Chapter Six identifies the attributes that promote the success of IT projects. Chapter Seven identifies the perspectives that promote the success of IT projects. Chapter Eight is the final chapter. It presents an overview of the discussions and the findings of the study. The limitation of the results and a discussion of the robustness checks for the models are also provided in this chapter. The conclusion restates the study's contribution to the existing knowledge and provides recommendations for further research areas.

ACKNOWLEDGMENTS

I always grateful to all the blessing and trials I received during writing this book. Both Prof. Kieth Hales and Iain Morrison whom I truly respect, are the important pillars of my foundation of knowledge. Prof. Dirk Hovorka, Prof. Gavin Finnie, Prof. Janet Price and Prof. Alan Finch, my ever close friends for providing the listening ear when there are situations without their contributions is impossible to achieve my ambition. My colleagues Prof. Wayne Irava, Dr. Pradeep, Dr. Kevin Tang, Safdar Khan, Marina Osman, Kay Imukuka, and Nigel Cartlidge, whom we used to play, eat, travel and share every single secret within on the research journey. Your sincerity to provide assistance whenever I need is truly overwhelming.

My warm thanks extend to my close friends and colleagues Dr. Emma Chavez- Mora, Dr. Sabina Cerimagic and Ingrid for sweetness and love. Their support was fully enjoyable, unforgettable and much appreciated, and in one way or another, provides me inner power and joy the field of work I chose. Their support in all events provides me with insights that nothing is impossible to achieve.

I also thank the Bond University Research Committees (BURCS) and the associated research committees. Being a High Degree Research (HDR) representative at the University for one year, President of HDR Academic Club for two years and Research faculty representative for one year gives me deep understanding regarding the research environment. All of these experiences enhance my own ability and awareness to complete this book.

Thanks to my friends in Oman: Lazina Dosormiers, Andrew H. Lukat, and my teacher Nasser Abdullah Al-kindi. My utmost gratitude for my family, children, and mum for being so patient during my working with this book. Their patience in training and guiding me enables me to finish this book. I send this academic achievement as a gift to them and I hope it will serve an inspiration and add strengths in achieving their own goals. 'Knowledge is power' – go for it.

The above colleagues, friends, and families, etc contribute to the book you are looking right now and serve as my stepping stone in fulfilling my own adventure in IT professional career.

Dr. Abdullah Ali Tweir Al-Hatmi
Bed, MA, MBA, and PhD

1. INTRODUCTION

1.1 GENERAL REMARKS

The uncertainty of value generation from government investment in Information Technology (IT) and the associated problems that arise from the complexity of implementing IT projects within government agencies is a topic that deserves serious attention. This research addresses this issue, exploring the impact of strategic alignment on the success of IT projects in a public organisation.

Chapter One contains an overview of the entire book. Section 1.2 provides the research background. This is followed by Section 1.3, which explains the motivation and justification for undertaking this research; namely identifying and addressing the shortcomings that exist in current approaches to value realisation. The section also identifies how such discrepancies can be addressed to increase the probability of more 'IT efficient' organisations.

The research statement of the study is outlined in Section 1.4: *'To examine the aspects of strategic alignment and their impact on IT projects in an Australian local government context, recommending areas where business/ IT alignment can be improved'.*

Section 1.5 provides the definitions of the key terms used in the book. Section 1.6 identifies a research gap in the literature related to SA on government IT projects. Section 1.7 outlines a further research question: *'How should strategic alignment be deployed in an IT governance context to ensure the success of government IT projects?'.* Section 1.8 contains the research methodology used in this study and also reviews the research approaches in case study analysis and the purpose of enquiry. Section 1.9 describes the case study undertaken and the scope of the research. Section 1.10 addresses the contribution of the research, and the last section, Section 1.11, summarises the whole book.

1.2 RESEARCH BACKGROUND

Governments around the world have followed the lead of the private sector in integrating IT into their processes and the interactions among staff, for example introducing email distribution lists for their constituents and developing intranets for employees to communicate and share information. Both UNESCO and the World Bank actively encourage and assist governments in applying IT to their various public agencies. For example, the UNESCO Institute of Information Technologies in Education (IITE) supports reinforcement of national capacities in ICT policy development for education of people with disabilities (Kotsik and Tokareva 2007). This program has offered the specialised training course of ICTs in education for people with disabilities in many countries. Similarly, the World Bank has highlighted the role of information, knowledge and applying ICT for international development (ICT4D). The ICT4D program is concerned with using ICT to help socioeconomic development and human rights and, in the process, bridging the technological gap between First World and developing countries. By encouraging the use of Internet-connected computers in public organisations, the ICT4D program requires new technologies, new approaches to innovation, intellectual integration and a new view of the world's poor (Heeks 2008).

Therefore, most public sector organisations (PSOs) depend on Information Technology (IT) as an essential part of their day-to-day operations. As IT has become more powerful, its use has spread throughout organisations at a rapid rate. Different levels in the management hierarchy are now using IT, where once its sole domain was at the operational level. For example, the sharing of knowledge and accountability about decision-making on government IT expenditure amongst stakeholders has become an important part of public policy to emphasise corporate social responsibility.

However, as in business, the results of many government ICT initiatives have been mixed. The emerging consensus is that the concentration on technological solutions rather than processes has not yielded the desired results and here is currently a high rate of IT project failures in public sector organisations. There is of serious concern (NASCIO 2007; Lobur 2011) as public organisations allocate considerable resources to government IT/IS (Information Systems) project initiatives (Cresswell and Burke 2006). The failure of IT projects has distinct and grave effects on any organisation's prospects. If the IT system introduced involved the commitment of substantial resources and funds, project failure can result in the loss of invested capital as well as forgoing the benefits that would have be available had the project been successful(Al-Hatmi and Hales 2011).

Moreover, as managerial tasks become more complex, the nature of the required information systems (IS) changes, from structured, reutilised IS support to ad hoc, unstructured, complex enquiries at the highest levels of management (Galliers and Leidner 2003). Thus, relating IT initiatives to government strategic objectives remains a multi-agency challenge for ICT and non-ICT executives alike. Emerging research indicates that significant 'across the board' gains can be achieved when ICT initiatives and organisational strategies are successfully aligned through cross-boundary collaboration (Elpez and Fink 2006; NASCIO 2007). However, there has been little discussion to empirically investigate the relationship between SA and the IT project success.

Through strategic alignment, the aim of IT integration now is not only to improve efficiency but also to improve business effectiveness and to manage organisations more strategically. Hence, based on organisational goals, ICT values are considered essential for any IT project implementation. Given the complexity of government structures, managing public technology is considered to be increasingly important (Grembergen 2004; De Haes 2007; Luftman 2011). SA can improve organisational performance and the delivery of government public services(Guldentops 2004; Luftman 2011) that ensure all the anticipated benefits are achieved (The World Bank 2007).This can be realised through the participation of the IT and business managers, well-defined IT/business strategy and the prioritisation of IT investments.

1.3 RESEARCH MOTIVATION

IT/IS spending by governments is huge and increasing. IT research and advisory company Gartner estimated that total spending on IT globally would reach approximately to U.S.$2.53 trillion in 2006 (De Souza, Nariwawa et al. 2003). In recent study by Kilkelly (2011) however, the study shows that the total annual cost worldwide IT project failure alone is $6.2 trillion dollars according to an IT-industry expert Roger Sessions. A 2002 Gartner survey found that 20 per cent of all expenditures on IT is wasted (spent on ineffective systems with no added value IT) which represents, on a global basis, an annual destruction of value totalling approximately U.S.$600 billion (Huber 2002). Similarly, the US federal government spend approximately $76bn on IT in 2009 (Gross, 2009). The US government also identified approximately 413 IT projects, cost around $25.2 billion in expenditure, as being poorly planned or performing (Powner, 2008).

The Ministry of Work and Pensions in United Kingdom (UK) wasted more than two billion pounds by abandoning three major projects since 2000 (Guardian 2008) and the net cost of a gun registry in Canadian government is 500 times the original estimate, with IT representing over 25 per cent of that cost (Canada 2007). In a daily UK mail (2012), the UK's National Health Service (NHS) project was financed to digitize patient records in addition to linking all of the different parts of <u>the UK's NHS</u>, but deemed to failure and costed approximately $20.1 billion, the largest IT project failure of its kind in the world. The lost money could be enough to pay the salaries of more than 60,000 nurses for a decade.

Likewise, the Australian Federal Government invested heavily in IT/IS with programs such as Networking the Nation (AU$77 million), Building on IT Strengths (AU$2.9 billion), and Backing Australia's Ability (AU$464 million) (ALIA 2003). Overall, Australian organisations spent AU$20 billion on IT services in 2003 (Gartner 2004). The failure of Australian Customs Service Systems project alone cost millions dollars and its effects extended to importers, exporters, and intensification disorder amongst social groups (Marshall, 2006).

Top management have found it increasingly difficult to justify rising government expenditures on IT projects and IS (Counihan, Finnegan et al. 2002) and are often under immense pressure to find reliable measurements for the contribution of their organisation's IT/IS investments to business performance that ensure the anticipated benefits are actually realised (Lin, Pervan et al. 2005; ITGI 2008; Dodd, Yu et al. 2009). The difficulties in measuring benefits and costs are often the cause for uncertainty about the anticipated sometimes intangible public benefits of IT investments and hence are the major constraints to IT investments. Moreover, the realisation of expected benefits is a complex process, as it involves many integrated factors including financial and organisational plans, processes implementation, stakeholder involvement and technical requirements which are often dealt with ineffectively by organisations (Mirtidis and Serafeimidis 1994; Seddon, Graeser et al. 2002; Thomas and Mullaly 2007; Crawford 2009), and particularly in terms of IT governance (Pervan 1998; Warland 2005). One of the challenges faced in Australia, as noted by Matthew Boon, Managing Vice President at Gartner: "Many organisations in Australia are running out of space in their data centres, coupled with fairly old and inefficient infrastructure" (Gartner 2011, p.1). Some of the reasons why the benefits from IT investments are not being realised in Australian public organisations are:

- ***Poor definition of benefits:*** Due the nature of the public returns (for example, if benefits are unstable or intangible) and the diversity of input from public stakeholders, benefits are poorly defined and little attention is given to the intangible benefits when decisions are made(Lin, Pervan et al. 2005);

- ***Insufficient measurement by traditional financially methods (for example, ROI and NPV):***These methods are insufficient in terms of quantifying the relevant benefits and costs in public organisations, such as improved investigative quality and a safer, more secure community (ITGI 2007);

- ***Insufficient knowledge-sharing:*** Many IT project failures are due to one or more of the following reasons: insufficient awareness of organisational issues; insufficient involvement of stakeholders in the public organisation; inadequate training of users; and poor alignment of IT adoption with the business strategy (Krauth 1999; Elpez and Fink 2006; Wang and Belardo 2009).

- ***Lack of business/IT alignment:*** This reduces the potential value of IT investments(Firth, Mellor et al. 2008);

- ***Excessive cost:*** It is too costly to undertake the proper business case (BC) and post-implementation reviews (PIR); and

- ***Immature of IT governance:*** There is lack of IT connectivity and integration of the different parts of the organisation that affect adapting government online initiatives (Guthrie 1997; Huang, D'Ambra et al. 2002;Wilkin and Riddett 2009).

Despite the focus on IT project challenges by partitioners and researchers, there has been relatively little attention given to the success and failure of individual IT projects(Standing, Guilfoyle et al. 2006), or even how to adequately measure the success and failure of projects. Project success can be classified by whether the anticipated benefits were delivered within time and budget constraints, but the difficulty arises if the benefits were not clearly defined or cannot be measured effectively. To explore these issues, this book examines how SA attributes and perspectives contribute to success in relation to IT projects and explains how this information can be used to improve the outcome of projects.

SA places a strong emphasis on value delivery and is therefore of interest to governments globally, as approaches to SA normally used in private organisations are not always effective in public sector organisations (PSOs) due to the different standards of value accountability and measurement used in government and commercial organisations. Genuine efficiencies and service improvements will only occur if governments and their agencies adopt a holistic approach to the SA perspectives and attributes identified as important in public sector organisations. A holistic approach emphasises the processes governing the entire ICT life cycle and its management (Lainhart 2008).

There is a need for continuing research on strategic alignment in the public sector that examines the SA perspectives associated with IT governance and their relationship with IT value delivery (Green and Ali 2007) and project success. Two of the most commonly discussed examples of deliverables in public sector organisations are efficiency and effectiveness; however, other benefits can be measured and assessed, such as community safety, sharing knowledge between government and citizens and systematic ICT decisions between stakeholders and the public, and improving the quality of public services. These can be achieved by establishing cross-boundary cooperation between all of the stakeholders in public organisations and understanding and understanding the special need for communication between IT and business managers (Cresswell, Burke et al. 2006; Mocnic 2010).

Therefore, the motivation for this book is to better understand the failure of government IT projects. To achieve this, an analytical framework was developed and used in the case study of Council, a local government. Based on the results of the research, recommendations and conclusions were developed that will provide the tools to promote the guidance of SA measurement and, therefore, increase the probability of the achievement of business and IT alignment success for business practitioners and decision-makers.

The use of SA perspectives in IT projects is recommended as it can assist with governance and decision rights concerning the use of IT (Weill and Ross 2004; NASCIO 2007, Weilbach and Byrne 2009), which in turn increases the chance of project success (Byrd, Lewis et al. 2005; Verner and Evanco 2005; Gartlan and Shanks 2007).

1.4 RESEARCH STATEMENT

Numerous IT projects have failed to meet expectations (Whittaker 1999; Lobur 2011). This can be due to a lack of clear definition and understanding of the benefits of the IT investment (Schniederjans and Hamaker 2003) before management commitment is made and funding approval is provided (Jung 1999; Stewart 2008). This failure to appropriately strategically align the planning of IT investments generally results from a limited understanding of the methods, skills and tools required for selecting the portfolio of IT projects which adds the greatest IT value to the public sector organisation. This understanding can be enhanced through applying the appropriate perspectives and attributes of strategic alignment (Elpez and Fink 2006; Pardo and Dadayan 2006; Firth, Mellor et al. 2008; Al-Hatmi and Hales 2010). Therefore, the purpose of this research is:

To examine the aspects of strategic alignment and their impact on IT projects in an Australian local government context, recommending areas where business-IT alignment can be improved.

Strategic alignment is generally defined as selecting appropriate alignment perspectives for achieving business objectives (Henderson and Venkatraman 1993). Therefore, the outcome of this research is aimed at assisting senior management to identify the appropriate framework when making decisions in their organisation and to ensure feedback on value generation assists in providing governance assurance.

1.5 DEFINITION OF TERMS

The terms involved in this research are set out below.

1.5.1 STRATEGIC ALIGNMENT[1]

'Strategic alignment' (SA), 'coordination and linkage' and 'strategic fit' (Lederer and Mendelow 1989; Henderson and Venkatraman 1993) have all been used to describe the concept of alignment (De Haes 2007). The term 'strategic alignment' is used in this research and its definition is adapted from Luftman (2000) to mean applying IT in an appropriate and timely way in harmony with business, strategy and needs.

This definition is in line with the dynamic processes of a project's life cycle not traditionally understood as 'an outcome' (Reich and Benbasat 1996) that do not focus on IT plans and social aspects (Reich and Benbasat 2000), such as communication and planning. These processes instead focus on applying IT (Shimizu, de Carvalho et al. 2005) for delivering value to the public (Hales 2005; Cresswell, Burke et al. 2006; Green and Ali 2007; Lobur 2011).

1.5.2 STRATEGIC ALIGNMENT PERSPECTIVES

The term 'perspectives' here is used to capture the same previous concepts identified by literature, such as 'factors' (Gartlan and Shanks 2007), 'quadrants' (McFarlan 1995), 'enablers or inhibitors' (Luftman 2000), and 'principles' (Bhansali 2007).

1.5.3 STRATEGIC ALIGNMENT ATTRIBUTES

The term 'attributes' is used to refer to specific characteristics of each perspective.

1.5.4 PROJECT MANAGEMENT

Since a local government provides many services to the public at large, the concept of project management when applied to local government exists in a broader context that includes program management and portfolio management. There can be a hierarchy of strategic plans, portfolios, programs, projects and subprojects in which a program consisting of several projects contributes to the achievement of an overall strategic plan. Thus project management is defined as the application of knowledge, skills, tools and techniques to project activities to meet project requirements. Project management is accomplished through the application and integration of project management processes. The project manager is the person responsible for accomplishing the project objectives.

1.5.5 IT PROJECT MANAGEMENT

The term 'IT project management' is used in this study to mean the implementation of an IT project master plan for the recommended solution with details, costs and benefits as well as specific timing and resource requirements. In short, it is defined as a set of methods, processes and practices that are repeatedly carried out to deliver IT projects (Council 2007).

In the case study analysis, the researcher examined documents related to the project including the

---

1   The terms 'alignment' and strategic alignment' are often used interchangeably. However, using 'strategic' does mean alignment at a higher level, and since the research context (IT governance) of this research focuses on the top level of management, 'strategic alignment' is used in this book.

Business Case (BC), Project Management Plan (PMP), Charter, Value Realisation Plan (VRP) and Post-Implementation Review (PIR). Hence, it is necessary to identify management practices and define IT project management.

1.5.6 IT PROJECT

A project is a temporary endeavour with a defined beginning and end (usually constrained by date, but can be by funding or deliverables), undertaken to meet unique goals and objectives, usually to bring about beneficial change or added value(Council 2007).

The term 'IT project' means a project which is primarily directed towards the development or improvement of an IT application(s) or IT infrastructure (Schwalbe 2006). IT projects, therefore, can be very diverse and not restricted to the more traditional view of software development projects (Stevens 2011).

1.5.7 IT PROJECT SUCCESS

In this study, 'success' in an IT project is used to refer to four critical success factors identified and supported by Ambler (2007); McNamara (2005); and (ITGI 2008): budget, time, scope and value realisation (achieving benefits). These terms are discussed in greater detail in Sections 4.4.1 and 4.4.2

I think success is really time, budget and realising the benefits. Most importantly, realising the benefits. They are critical to the value proposition as to why you started the project in the first place... So you think about the outcomes - that is really what is important. (Participant 8)

1.5.8 IT GOVERNANCE

'IT governance' is defined as the processes whereby SA is deployed and examined, and is the research context for this study. It consists of the organisational structures, processes and relational mechanisms that try to ensure that an enterprise's IT sustains and extends the organisation's strategies and objectives (ITGI 2007). In this research, the definition that captures the concepts of structures, processes and relationship mechanisms relevant to government context is adopted from NASCIO (NASCIO 2008, p.2), which defines IT governance as:

A condition in which government is effectively using information technology in all lines of business and leveraging capabilities across state or government appropriately to not only avoid unnecessary or redundant investments, but to enhance appropriate cross-boundary interoperability.

In this context, 'government' refers to structures, 'all lines of business' refers to processes and 'across state' refers to the relationship mechanisms in an organisation. It is in the organisation's best interest to clearly define who makes the tough decisions and who is answerable for failure or praised when profits are made and benefits are met. A key issue in SA perspectives is who makes the systematic decisions on government IT expenditure and the calculation of the returns in IT investment. Therefore, this research investigates the impact of SA perspectives on government IT projects.

1.5.9 PUBLIC VALUE OF IT INVESTMENTS

Value is not a simple concept; it is complex, dynamic and involves numerous stakeholders (Hales 2005),both internal and external. For public sector organisations, value is more complex than for private sector organisations in that it is often non-financial in nature (ITGI 2006). In this research, public value refers to the benefits to organisations and the public and encompasses both financial and non-financial benefits. Such benefits include efficiency gains, interoperability, increased capability, increased effectiveness in operational processes, avoided cost and improved customer services and customer satisfaction. Economic/financial metrics are sometimes used to measure the underlying aspects of public value in financial term such as Return On Investment (ROI), net profit and bankable financial benefits.

1.5.10 MATURITY LEVEL

The criteria used to measure the maturity levels of the focus areas of the SA perspectives are simple but effective. The five SA perspectives areas outlined below:

- **Strategy**: to ensure that objectives are clear, measurements are monitored and the IT/business plan has adequate and accurate details
- **Knowledge**: to ensure that responsibilities are shared and knowledge spread across organisations, that there is communication between IT/business managers and that there is cross-training and a well-communicated information flow.
- **Decision-making**: to define how and by whom systematic decisions are made concerning IT investments and budgets, including the involvement of stakeholders in decision-making.
- **Enterprise architecture**: to integrate business/IT solutions, minimise risks and ensure the capability of applications and technology.
- **Public value**: to ensure the improvement of customer services, efficiency gains, avoid unnecessary cost and improve economic and financial metrics.

For each of these perspective areas, the maturity model is classified into five levels (see also Table 4-7) starting from the initial 'ad hoc' stage, which indicates the lowest level of maturity, to the 'optimised' stage, which indicates the highest level. The five maturity levels are:

- **Initial/ad hoc:** business/IT alignment is low or non-existent
- **Committed:** there is commitment for business to become aligned with IT
- **Established:** business/IT alignment is established and focused on business objectives
- **Improved:** SA of IT is centralised as a value asset
- **Optimised:** business/IT alignment is highly integrated in the strategic plan and has reached a co-adaptive stage

1.6 RESEARCH GAP

The gap between government expenditure on IT investment and the realised value of IT in public organisations is widely discussed by practitioners and researchers in the IT community (Guldentops 2004; NASCIO 2007). Despite this focus, IT projects continue to fail in public organisations(Byrd, Lewis et al. 2005; Pardo and Dadayan 2006). Improved strategic business/IT alignment is needed (Motjolopane and Brown 2004; Luftman 2011) to address the repeated mistakes which cause this continued failure, such as poor decision-making when selecting which IT project(s) to invest in (ITGI 2005) or inappropriate decisions made during the project implementation process.

Without proper alignment of IT, it is unlikely that any enterprise will achieve and sustain long-term success through the delivery of value to its stakeholders.(ITGI 2005, p. 7)

Historically, the models for measuring value attributes are simple and financial and relate to cost reduction or revenue. However, these models are more suitable for private companies where value is usually measured in profit. Public sector organisations generally have more complex governance constraints and value perspectives. Earlier strategic alignment models do not accommodate the specific challenges facing governments caused by the rapid change of technology in a dynamic environment (Weill and Ross 2004) and the value vector from existing models is insufficiently complex when aspects of public value are to be taken into account. Even the various parameters used by governments to evaluate their IT projects, such as Economic Value Added (EVA) and Return on Investment (ROI), are inadequate tools to measure the public value of IT investments in terms of speed, accessibility, efficiency and the political goals of government services. Additionally, due to poor management and insufficient checks and balances in IT investment activities, IT projects can cost more than twice their original estimate (Schniederjans and Hamaker 2003), further decreasing their apparent value. So, what earlier models measure is inadequate and what appears critical to success has no equivalent and concrete framework to determine which of the strategic alignment perspectives contributes to the success (Bhansali 2007).

Although the use of technologies by public sector organisations has the potential to improve the lives of millions of people around the world, there is still currently only sparse and largely anecdotal research investigating the role of SA on IT project success in public organisations (ITGI 2008). Adopting an effective SA model is essential for governments to increase the chance of IT project success. This research, therefore, proposes a conceptual framework of SA to investigate how strategic alignment perspectives are deployed in an IT governance context to ensure the success of government IT projects.

1.7 RESEARCH QUESTION

The main Research Question (RQ) is then developed based on the research problem identified.

RQ: How should strategic alignment be deployed in an IT governance context to ensure the success of government IT projects?

1.8 RESEARCH METHODOLOGY

The research paradigm adopted for this research is an interpretive qualitative case study approach (Stake 1995; Yin 2003). Qualitative approaches are vital for theory-building in emerging fields (Denzin and Lincoln 1994), such as strategic alignment in public sector organisation, as they allow for the use of multiple research techniques which can enrich understanding of challenges and opportunities for delivering IT value to the public (Dawes, Burke et al. 2006). This research involves the study of IT projects in a local government context in Australia. The case study approach is particularly well suited to address the research question, as the case study provides a public organisational context (IT governance) for the study of the relationship between the SA maturity level of IT projects and their success while also allowing for the identification of additional measures of alignment, for example organisational capability, and their success rate. Such an approach will add to the existing body of knowledge in an integrated way that identifies and analyses the SA perspectives and attributes needed to provide a continual management process of the life cycle of IT project. It also recognises the complexity of the research question and allows for the assessment of SA perspectives in a natural, unaltered setting. The case study explores in depth the organisational complexities and life cycle processes of IT projects and the alignment of the processes to the strategies and goals of the organisation.

The methodology of this research was based on three approaches. Firstly, a review of archival records was undertaken. These documents provided a rich history of the development of IT projects over time. The evolution of IT is clearly depicted in these documents. Archival sources note that the scope of IT is expanding and question its role in the future, which is interesting because it shows when questions and issues of implementing information systems becomes a core of IT governance practices. As a result, a rich history of the process involved in IT projects over time was obtained.

Secondly, participatory observation was carried out during data collection after IT projects had been implemented. This method was important in obtaining personal opinions from staff who were not the business/strategy owners. IT management practices in the local government were observed and the researcher studied how effectively employees communicated was studied whenever the opportunity was given during data collection, training or during an informal coffee meetings (see also Section 3.5.2.1).

Thirdly, a series of semi-structured interviews were conducted with eleven knowledgeable senior IT executives in local government staff over a one-month period. Each interview took between one and two hours and participation was voluntary. Interview responses were kept anonymous in this book, with each participant being assigned a number for identification. The interviews involved questioning former and current employees from different management levels. The data obtained provided an insight into understanding and explaining SA in the context of a local government. The data was analysed using qualitative research methods and the results used to answer the study's research design and questions.

The triangulation of these approaches above in qualitative research allows for a greater understanding of multiple realities as well as addressing the tacit and explicit aspects of SA that generate value from IT projects. The adoption of a triangulation (Denzin 1978) of data collection methods and sources within the research strengthens both the credibility and validity of the research (Irava 2009). The data compiled was then analysed using the NVivo tool and according to the research design and themes: IT governance; strategic alignment perspectives; and IT projects. This software was used to code, manage and analyse the data. As this book is using qualitative interpretive research, no statistical tests were used.

Figure 1 illustrates the organisational settings for ICT strategy where Strategic Alignment was examined in the IT governance and context process. The IT governance structures were categorised by their roles; as decision-makers, process owners[2], knowledgeable about ICT governance practices, responsibility for IT investment and stakeholders in various business units, including the process owners of the ICT Portfolio[3]. Such structures were populated by the local government participants in this research.

---

[2]   Process refers to the overall management of BC (Business Case), PMP (Project Management Plan) and PIR: (Post-Implementation Review)

[3]   Information and Communication Technology (ICT) Portfolio consists of CPP (Corporate Performance and Planning), BG (Business Group), PMO (Project Management Office), CIO (Office of Chief Information Officer) and Ops (Operations)

FIGURE 1-1: ORGANISATIONAL SETTINGS

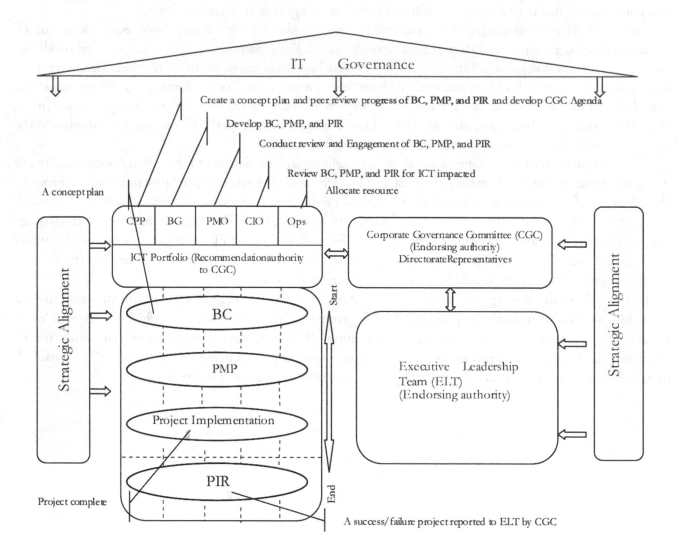

Ethical issues should not be raised during data collection (Creswell 2009). If they are raised, they could affect the data. Therefore, precautions were taken on ethical matters to protect the participants and their settings. Validation of qualitative research occurred throughout the whole research process, with construct validity, internal validity, external validity and reliability taken into account. As most public sector organisations run numerous IT projects and the failure of government IT projects is not often publicly admitted, a single case study context provides an appropriate methodological opportunity to study the phenomenon where it occurs.

1.9 RESEARCH SCOPE

The research undertaken was conducted within a set of boundary limitations. An administrative local government in Australia was chosen as the organisation in which the study would take place. Australia has a three-tiered system of government, which different levels of government operating at a federal, state and local level. The local level is called a Council. The Council where the case study was taken is the second largest council in Australia with more than 3000 employees. It appears to have established a sound IT governance (ITG) framework in its daily operations. The term 'governance' is essentially associated with accountability and responsibilities within an organisation that pay particular attention to organisational structure, management mechanisms, and policies (OGC 2008).

A total of eleven IT professional participants were selected to be interviewed, thus the study followed a purposeful sampling approach. These participants were characterised by being a top-level core owner source of IT governance and practices and therefore provided direct and rich information about the experiences and the issues they encounter in government IT initiatives. The characteristics of the IT professional sample are explained in Section 3.5.1.

The fourteen projects analysed as part of this study were randomly selected and vary from small (AU$500,000) to large project (AU$2m) and they all occurred between 2004 and 2009. They are all considered pure IT projects and are initiatives that are implemented into Information Communication and Technology (ICT) in a local government environment. The details of these projects are given in Section 4.3.3. Since SA cannot be studied in isolation, the IT governance context in the Council is a relevant one in which to study explanations for project success because:

- It consists of project documentation concerning the project being undertaken such as Business Case (BC), Project Management Plan (PMP) and Post-Implementation Review (PIR);
- It consists of a hierarchy of sharing decision-making for IT investments;
- It has an ideal complexity of government IT environment (for example, the funding of projects is often unstable); and
- It shows that projects do 'fail' and unnecessary costs still occur; and
- It allows for access to information about IT projects that is often inaccessible publicly due to political and security reasons.

All ethical procedures were considered before the data collection took place. For example, a gatekeeper letter from the Council was provided to the researcher allowing him to conduct research (Appendix A). In addition, a consent form was provided to each participant before their interview which stated the participant's right to withdraw at any time without given reason and explaining that they will be kept anonymous in the book so that they cannot be identified in any way (See Appendix C). The details of the ethical procedures used in this research are outlined in Section 3.6.

Finally, observation, archival research and semi-structured interviews were used to collect data in this case study. The use of various sources involved providing full descriptions of the research problem addressed through project documentation such as BC, PMP, and PIR, meeting minutes and participant inputs. A description of the methods used in data collection is provided in Section 3.5.2.

1.10 CONTRIBUTION

The research outcomes reported in this book contribute to the body of knowledge in a number of ways. The major contribution of this book is an original investigation within an Australian public sector context on the impacts of augmented SA perspectives on the success of IT projects. The results will identify how perspectives are deployed in an IT governance context where the success or failure of government IT projects occurs. This, in turn, provides a greater understanding of how to generate public value through IT projects. In doing so, the study will contribute to practice and theory and address a current gap in the research and literature.

1.10.1 CONTRIBUTION TO PRACTICE

Demonstrating the importance of understanding, designing and applying SA perspectives in government agencies is one of the main contributions of this research. Researchers, regulators and educators can now design, prioritise and apply these perspectives using a holistic approach.

The study provides guidance on how practitioners and government policy-makers can maximise the value of their IT investments and reduce the failure rate of IT projects by implementing strategic alignment perspectives appropriate to the context to their organisational processes. It is anticipated that an improved alignment of local government strategy and ICT strategy will contribute to greater success of government IT projects and provide guidance and direction to governments to optimise the value of their IT investments. The model developed in the book provides both insight into the contributing aspects of project success and a tool for professionals and government policy-makers to apply in order to achieve strategic alignment when implementing government IT projects. This will facilitate their project implementation in order to ensure that the projects achieve the strategic goals and business objectives of the organisation.

A practical model of how to achieve strategic alignment could result in significantly improved government IT projects with a greater probability of project success and, as a result, enhanced efficiency and effectiveness in public services.

1.10.2 CONTRIBUTION TO THEORY

Based on a clear conceptual framework, this research expects to generate knowledge which will further enhance the development of SA models and theories.

The contribution of this research to theory firstly commences with an input/output analysis of the role of SA and the construction of a maturity model for IT. This research investigates how SA perspective maturity levels influence the project outcomes, thus endeavouring to create a clear linkage between the maturity levels of the projects and their outcomes that are also identified and established by IT project success criteria. By investigating this linkage and other various relationship patterns, the researcher was able to identify which SA perspectives and attributes are more significant in ensuring project success. The research, therefore, theoretically advances the understanding of gaining public IT value through the successful adoption of IT projects. Hence, a conceptual model was developed.

Secondly, the case study analysed in this research demonstrates that this approach is a useful means by which the aspects of SA perspectives can be understood. This study also contributes to the body of knowledge in terms of the considerable methodological guidance provided on how a case study of the influence of SA on IT projects is undertaken. The selection of the case study, the participants and the data collected were all identified as relevant to the research. The knowledge gained from the interviews with the IT/business managers, IT governance staff and the business users of the systems in the case study provided insights to the research question and yielded useful answers.

The research contributes to the theories of SA by synthesising diverse literature on SA and IT projects and providing a set of comprehensive perspectives and attributes in alignment that can be used to uncover and address alignment issues during the life cycle of projects to create collaboration between business and IT strategy that may have otherwise remained untouched.

The book is structured as follows. Chapter Two contains a literature review on the background of IT, business and governance theory. It discusses strategic alignment perspectives and models and, by identifying gaps in the research, isolates areas that require further analysis in the context of public organisations. The chapter concludes with a suggested conceptual framework to be adapted to a government context.

Chapter Three describes the research design and case study methodology used to investigate the impact of SA perspectives on government IT projects in local governments. The justification for adopting a case study methodology is provided. Sample data collection and analysis procedures are given in detail in this chapter, providing the rationale concerning the validity and reliability of the methods used followed by the operationalisation of the theoretical constructs.

Chapter Four describes the organisational setting of the case study used in this research; a local government. This includes an overview of the organisation, the participants' characteristics, case summary characteristics and the characteristics of the IT projects ongoing in the local government. It also shows how the data was analysed and presented.

Chapter Five introduces the concept of SA maturity levels and discusses this concept in relation to the maturity levels of IT projects. This chapter also examines the success rate of IT projects and the relationship of success to the maturity level of IT projects. Thus, this chapter covers the main analysis, discussion and contribution of this research. Analytical analysis was undertaken further using the NVivo package tool, which helped to manage and analyse data within and across projects.

Chapter Six identifies the attributes that promote the success of IT projects. Chapter Seven identifies the perspectives that promote the success of IT projects. Chapter Eight is the final chapter. It presents an overview of the discussions and the findings of the study. The limitation of the results and a discussion of the robustness checks for the models are also provided in this chapter. The conclusion restates the study's contribution to the existing knowledge and provides recommendations for further research areas.

1.11 CONCLUSION

This chapter provided an overview of the importance of strategic alignment and how its perspectives are expected to increase the performance of IT governance, leading to the increased success of government IT projects. The research statement of this book was also provided. The research statement gave a clear purpose for the main topic of the book: an analysis of the impact of SA on IT projects in a public sector organisation. This chapter also discussed the motivation for this research as well as its objectives and the problem and limitations of the research. The gap in the research literature concerning the continuous failure of government IT projects was also discussed in this chapter.

A case study approach was used, as it is the methodology best able to provide the data that could fill the gaps in the research. The selected methodology, comprising of a case study, interview and observations, was explained and the outcomes they provided were questioned. The last section of Chapter One provided a brief overview of the contribution to knowledge and an outline of the layout of book. Concept maps describing this book are found in Figures 1-1, 3-1, and 4-1.

2. LITERATURE REVIEW

2.1 GENERAL REMARKS

In order to investigate the various aspects involved in strategic alignment, a thorough literature review of grounded theories has been conducted with a central focus on the impact of SA perspectives on IT projects. Hence, the purpose of this chapter is to provide an overview of the current understanding of the deployment of SA perspectives in government IT projects, identify the gap in the current understanding and state the research problem which addresses this gap.

This section contains general remarks on the chapter. Section 2.2 provides a historical background of the concept of alignment and a discussion of the current concept of SA in public organisations. Section 2.3 examines the notion of strategic alignment, the research context for this book and provides critiques of the development of theories and research on SA models. The section also discusses the SA frameworks that are available to business leaders in order to achieve a successful alignment between IT and business strategies. SA is an ongoing concern for many executives (Sledgianowski and Luftman 2005; Chan and Reich 2007), however because alignment problems are caused by a number of different factors, no one single framework has been developed than can solve all of the management issues. Sections 2.4 and 2.5 highlight the major flaws in the current models, including both the internal gaps that relate to the scope of applicability and the external gaps that relate to the inability of the model to meet the scope required to fulfil the complex application required. These include increased stakeholder accountability and the higher level of integrated systems common in public sector organisations. Section 2.6 provides a summary of the gaps in SA models. Based on the identified problems and the diverse ways in which the value of IT is returned in public organisations, the research question is refined, and the revised question is posed in Section 2.7. The current themes of a SA conceptual framework are also provided in Sections 2.8 and 2.9 to address the research question. The conceptual framework is used to facilitate a systematic linkage between business and IT strategies. This model is then developed throughout the book to demonstrate how it might be deployed in a government context to enhance a more suitable decision support framework that leads to increased project success by using the proper implementation process. The concluding section summarises the contents of this chapter.

2.2 HISTORICAL BACKGROUND OF THE CONCEPT OF ALIGNMENT

2.2.1 DEFINING STRATEGIC ALIGNMENT (SA)

The development of the concept of strategic alignment begins with the matching of business processes with innovation and technological systems. This collaboration of ideas promotes the idea that performance within an organisation can be enhanced by aligning the objectives of the organisational processes with the capabilities of the information technologies. Middleton defined alignment within an organisation as the internal practices of different aspects of the organisational efforts that focus on enhancing change and innovation (Middleton 2004). This includes pay (compensation), staff selection, staff promotion, staff retention and an organisational structure that can positively support organisational efforts to facilitate changes and foster innovation. However, in a joint study on how innovation and improvement actually happen, Robinson and Stern (1997) considered alignment as the degree to which the interest and practices of the employees are used to achieve the key goals of the organisation. The study of Robinson and Stern (1997)draws from the previous study of Reich and Benbasat (1996), which proposes the idea that alignment is linked to the degree by which information technology's mission, objectives, and plan support are supported by and are aligned with the business's mission, objectives, and plans.

Through the continuous study of alignment within organisations, the definition of the strategic models as well as approaches was developed. As seen in the works of Nelson and Cooprider (1996), Subramani, Henderson et al. (1999) and Tallon and Kraemer (2003), there are two approaches to alignment. These are referred to as the social and intellectual dimensions of alignment, whereby having good strategies, structure and planning methodologies in place are an essential antecedent to attaining a high level of strategic alignment. While the social dimension of alignment refers to a mutual commitment to both the business and IT missions from a human resources—managerial perspective (Reich and Benbasat 1996, p.58), the intellectual dimension of alignment examines strategic alignment at the cognitive level, i.e. the understanding shared between business and IT executives (Tan and Gallupe 2006; Walsh and Renaud 2010). In relation to strategic alignment, the approaches investigate the idea that there should be a well-defined state of high-quality functions in both IT and business objectives. If the two scopes are interrelated, a good plan using both dimensions can exist and alignment in the organisation can be achieved. However, it is important that the employees or the idea of the social dimension within the workplace should remain a vital consideration that clarifies the objectives, plans and mission of the business and IT (Reich and Benbasat 1996).

Using this view of strategic alignment as a starting point, the present generation of the business associates and IT experts continued their contribution to the field by exploring other definitions of alignment. There is no doubt that because of the rapidly changing nature of IT, the collaboration of the IT and business processes is also subject to change. For Luftman (2000), the collaboration of business and IT to form an alignment can only be appropriate once IT has become one with the business strategies, goals, and needs. Referencing Henderson and Venkatraman (1993),Luftman's 2000 study also emphasises the idea at within the strategic alignment, the selection of the appropriate perspectives should be considered in order to achieve the business objectives. Along with their contemporaries, Broadbent and Weill (1993) similarly agree that strategic alignment can be achieved as long as the business strategies are enabled; meaning that if the business strategies are defined and achievable, with the support of the IT system and strategies the functions and operations of the business are stimulated or enhanced.

Therefore, strategic alignment is a process that will help an organisation to secure clarity in their business direction, to have effective customer relationships, to have a resilient operating strategy, to have internal cohesion and collaboration, to adapt an organisational culture/infrastructure and to obtain sensitive leadership at all levels. Consequently, strategic alignment is a detailed process that encompasses a broad use of knowledge, technical ability and skills and the perseverance of all the members of the organisation, including the management and other stakeholders, to be able establish an extreme focus, realistic, flexible and continuous business infrastructure which is effective in meeting the needs of the business and satisfying consumers (Luftman and Brier 1999).

Strategic alignment also occurs when the Information Technology management performance merges with the essential strategies and core proficiency of the business organisation. When both of these are aligned, the capability of the Information Technology (IT) becomes consistent and amalgamated with the central strategic path of the organisation as a whole, which allows different stakeholders to create particular Information Technology-linked business forces and organisational strategic ways and directions. SA will reduce the risk and increase the efficacy and efficiency of an organisation, thus placing measurable business value on it (El Sawy, Malhotra et al. 1999). However, SA is about the linkage between people, organisation and process that supports IT capability and requires a relationship between business and IT to achieve the ultimate strategic goals of organisation. Having established the multidimensional nature of SA, the importance of SA and the consequences of misalignment are discussed next.

2.2.2 THE IMPORTANCE OF STRATEGIC ALIGNMENT (SA) AND THE CONSEQUENCE OF ITS MISALIGNMENT

2.2.2.1 THE IMPORTANCE OF STRATEGIC ALIGNMENT

The importance of alignment has been well documented in research literature (McLean and Soden 1977; Dixon and John 1989; Niederman, Brancheau et al. 1991;Chan and Reich 2007). It has been considered that a key benefit of having aligned IT and business strategies is improved decision-making concerning IT-related investments (ITGI 2005). Alignment grows in importance as organisations strive to link business and technology (Gartlan and Shanks 2007) in light of dynamic business strategies, continuously evolving technologies (Papp 1995) and high IT expenditure (Cresswell, Burke et al. 2006). The idea of strategic alignment has become the most pertinent and emergent issue within modern organisations, which requires business strategy to collaborate with the IT system and business processes (Tallon and Kraemer 2003). Many researchers argue that the most basic way in which an organisation can improve performance is to achieve harmony between the functions of the business objectives and the IT systems. This balance is attained by defining the strategic alignment (Sabharwal and Chan 2001). Researchers have found that firms often fail to recognise all of the possibilities that are available if business is interrelated with their use of IT supports, eventually leading to poor performance. This idea suggests that by not defining the strategic alignment, the collaborative potential of the business processes and practices and IT strategies may not be fully reached (Henderson and Venkatraman 1993; Pereira and Sousa 2005).

The role of strategic alignment within organisations has received much attention since the importance of IT/business integration was realised. Business leaders and IT developers sought out opportunities relating to strategic alignment, attempting to find the best ways to apply strategy to achieve better performance. Reich and Benbasat (2000) identified that the establishment of a strong alignment between information technology (IT) and organisational objectives has consistently been reported as one of the key concerns of IT/IS managers. They define alignment as the degree to which the information technology mission, objectives and plans are supported by the business mission, objectives, and plans. Since the late 1990s, the importance of IT in organisations has risen significantly, correlating with a huge growth in the sale of technology and increased day-to-day usage of IT-related services. Accordingly, much has been written in research literature on how to manage IT more effectively (Kaplan and Norton 1992; Henderson and Venkatraman 1993; Hodgkinson 1996; Galup, Dattero et al. 2007; Lynda M. Applegate, Robert D. Austin et al. 2007).

In the view of the increased importance of IT, Kearns and Sabharwal (2007) argue that in order to improve organisational performance, achieving strategic alignment between business and IT is essential. Organisations often fail to recognise the business potential that is related to their use of IT support, even though ignoring this potential can lead to poor operational performance. This idea suggests that in the absence of strategic alignment, the effective collaboration between business processes and practices and IT strategies may not be fully achieved(Henderson and Venkatraman 1993; Pereira and Sousa 2005). Sabharwal and Chan (2001) noticed this when they used Miles and Snow's typology to evaluate how the IS/ICT strategy alignment leads to superior business performance in different types of organisations. Croteau and Bergeron (2001) used the same typology and confirmed the importance of alignment between business strategy, technological deployment and organisational performance. In addition, CIOs have consistently considered alignment of IT with business strategy a top priority (Watson, Kelly et al. 1997). Thus, successful alignment between business and IT strategy is evident where both IT and business strategy demonstrate a planned alliance, which then leads to tangible, successful, business-focused outcomes (Gartlan and Shanks 2007).

Lee et al (2006) also found that to respond to uncertainties in global business environments, such as the agility of a globally distributed system development, organisations must not only develop global business strategies but also support them with information systems that are aligned with these strategies. Similarly, Singh et al (2007, p.59) stated that "information technology has significantly changed supply chain processes across different organisations and, in many cases, improved competitiveness". The improvement of the competitiveness and effectiveness of the supply chain can be achieved by aligning the use of technology with business transactions and thus facilitating the sharing of information and collaboration with suppliers and customers (Corsten and Kumar 2005).

IT has become an important aspect of everyday business and is vital to many business organisations (Gichoya 2005). For example, without the Internet, government agencies would not be able to offer online functions such as renewing passports, voting on government policy or in elections. The main driver which keeps management continuing to deploy technologies as part of the business processes is the idea that competitive position is somehow linked to the ability of the organisation to address and facilitate the key elements that are involved within the alignment of business and IT. To do that requires IT investment, however, investing in IT systems is not as easy as just deciding to do so. There are things that an organisation should consider before investing, for example: evaluating the accessibility of the IT systems; the policies and regulations that exists in the country in which the systems will operate, especially if the system is imported; and, most importantly, the ability of the system to answer the needs of the organisation. The decision-makers must understand that alignment is about creating harmony between the organisation and the technology itself. The more people who adopt a technology, the more opportunity designers have to create a more sophisticated product tailored to the industry. If the introduction of one technology is effective, the current trend is the acquisition of the same or more advanced technologies by rivals of the successful organisations, which gives the developers the opportunity to offer their products for a greater cost. Also, because of the continuous yet changing role that IT plays within various organisations across different industries, there are differences in the IT-adoption process that may increase the performance of the organisation and therefore eventually change importance of IT within an organisation. For instance, IT can automate the manufacturing process so that data feeds into the accounting processes, therefore becoming a part of the strategic resources such as enterprise resource planning (ERP) and customer relationship management (CRM)(Henderson and Venkatraman 1993).

The development of strategic alignment will therefore change if there are any alterations in the business objectives and IT systems. The utilisation of IT within an organisation requires the organisation to make IT a major factor in shaping the business strategies that enable it to perform effectively and respond quickly to the external environment of the business. As the literature indicates(Gartner 2003; Weerakkody, Janssen et al. 2007), measuring alignment can be difficult; however, the misalignment in between the business objectives and IT can be easily noticed.

2.2.2.2 THE CONSEQUENCE OF MISALIGNMENT

In recent decades, billions of dollars have been invested in information technology (IT) (Luftman and Brier 1999). The total worldwide expenditure on information technology (IT) exceeded one trillion US dollars per annum in 2001[4](Seddon, Graeser et al. 2002).

We have found the most difficult question to answer is to measure the effectiveness and the longer-term return on IT investment(IT Director in the health sector as cited by Seddon, Graeser et al. (2002, p.11).

---

4   1 Gartner (http://www.newsbytes.com/news/01/165194.html) estimated the IT services market in 2001 to be around U.S.$700B and growing at over 10% p.a. Adding computer and telecommunications hardware would take total worldwide IT expenditure to well over $1 trillion per annum.

PricewaterhouseCoopers remarked in 2003, as cited by ITGI (2005), that one of the most prominent IT-related issues that were being faced at that time was the perceived disconnect between IT strategy and business strategy. This lack of alignment leads to adverse business issues including:

- The inability of a business to reach its full potential;
- The failure to identify and capitalise upon business opportunities that could be enabled by IT;
- Potentially higher operating costs and, therefore, competitive disadvantage due to the failure to replace expensive labour-led processes with lower-cost (over the long term) automation;
- Incorrect and ineffective focusing of IT-related resources;
- The inability to recruit and retain high-quality IT and business personnel;
- Higher overall costs; and
- The erosion of stakeholder value over time.

Middleton (2004) examines the importance of organisational alignment for information systems success. He analysed the different arguments related to strategic alignment as seen in the studies of Collins and Porras (2004)and Reichheld (1996), both of which discuss perceptions regarding strategic alignment. He found that, overall, organisations did not appear to be aligned, with only 26 per cent of employees questioned feeling rewarded for innovations and 40 per cent not feeling integral to the organisation. Individuals may be willing to innovate, but since the organisation they work for is not aligned, then innovation and creativity suffer. Due to the relationship between business objectives and IT objectives, aligning the principles that run both into a combined IT/business plan can be an effective method to align the internal factors of an organisation, for example the interests, goals and motivations of the company and its employees, and thereby lead to greater company success and high profits. In their study, Henderson and Venkatraman (1993) also argue that the implementation of IT within the organisation cannot be effective without help from the business goals. The findings indicate the focus for information systems should move from the traditional management of software processes and technology in terms of time, budget and scope to organisational alignment, recognising that information systems projects are liable to fail if an organisation is not aligned. If an organisation is misaligned, this needs to be addressed before new information systems are implemented to enable it to succeed with information systems.

Without proper alignment of IT, it is unlikely that any enterprise will achieve and sustain long-term success through the delivery of value to its stakeholders.(ITGI 2005)

Robinson and Stern(1997) stated that alignment consists of mutual understanding between business and IT strategies, not only regarding their objectives but also concerning other components such as the cooperation of people or staff. Drawing from the work of Henderson and Venkatraman (1993), van Eck, Blanken et al.(2004) explained that there are three layers in strategic alignment that are working together: the infrastructure, the business systems and the business itself. These layers aim to ensure business and IT strategy are in harmony and to establish linkage between the operation of the IT infrastructure and business processes. However, the decisions concerning strategic alignment are not driven solely by business operations but by various forces; one of which is the business strategy. Because of the risks and uncertainties associated with the business strategy, alignment may not be achieved well with business system operations.

When we have gone back and done a post-mortem on many of our projects that did not turn out well, several things seem to always come up and poor alignment is one of those things.

An executive quoted in Griffith and Gibson Jr. (2001, p.76)

The consequences of failing to align the business strategy with the IT strategy have been observed by others in the literature.

For many years, IT alignment has been the No.1 issue on IT executives' minds. However, despite the focus and attention we've paid to this idea, we're no closer to IT alignment today than we were 20 years ago.(DeLisi 2005, p10)

2.2.3 THE CURRENT CONCEPT OF SA IN A PUBLIC ORGANISATION CONTEXT

In today's increasingly IT-driven culture, governments are expected to seize the opportunities offered by the technological advancements in IT. Linking business and IT strategy has been successful with modern commercial organisations (Grabski, S. V., S. A. Leech, et al., 2011), so the government and other public sector organisations should consider the principles and advantages of strategic alliance and apply strategic alignment in their use of the technology. Strategic alignment in public organisations can help to transform the works of the public organisation and strengthen their relationships with individuals, other businesses, the community and with the other public organisations. Part the purpose of strategic alignment is:

..to assist public sector organisations and governments in aligning their various organisational transformation initiatives with counterpart technology-related initiatives.(Ojo, Shareef et al. 2009p.i)

In a study of the public sector programs, public administration reform (PAR), in Macao, Ojo, Shareef and Janowski (2009) noted that two of the activities involved in strategic alignment include the assessment and alignment of the respective dimensions to achieve their objectives. There are two primary dimensions in which public sectors can improve in managing IT. The first is to enable technological support for organisational strategies and goals, and then secondly to facilitate the development of a technology-driven environment within the public organisations. The need for SA within the public sector is very evident, for example as with the application of the Electronic Government (eGOV), where the government's programs and technological strategy must be centralised in order to best utilise the IT and gain maximum benefits (Ojo, Shareef et al. (2009) and maximum a seamless interoperability standards for electronic government (Dos Santos & Reinhard, 2010).

When looking at the history of the process of IT and the context of SA within the government, it is evident that implementing SA is a challenging task for the public sector. The challenge is not only because public sector organisations invest a lot of money[5] in Management Information Systems (MIS) (Muniz, 2009), but even an innovation-led drop in annual per-gigabyte storage costs between 35-45%, has failed to halt the rise in storage spending which increases by average of 35% annually (Tallon, 2010). In addition, because of security and confidentiality, it is critical to design a system that can handle access authorization and authenticating for shared information across-boundaries (Lobur, 2011). Because of such reasons including organisational aspects, 'only one out of four development projects succeed. Projects are easy to start, but hard to finish' (Heiskanen, 2012 p. 1). It is also noted that high failure rate of IT projects is related to the poor management of risks during the IT project implementation (Carr, 1997; Stevens, 2011).

Therefore, it is vital for organisations to pay attention to issues of alignment in the planning stages as an important aspect before implementing IT strategies to promote an efficient use of IT investments. According to Al-Hatmi and Hales (2010), the benefits of IT investments and the values that the private sector organisations use tend to be more functional. In other words, to measure IT value and success they use parameters such as payback periods, Net Present Value (NPV), Internal Rate of Return (IRR), Economic Value Added (EVA) and Return on Investment (ROI). These parameters are used to measure organisational performance and profitability in financial terms but are less suited for evaluating other intangible benefits as required by public organisations, such as ensuring a healthy and safe community or measuring the quality and effectiveness of public services.

IT structures in private organisations are modelled and developed through a series of the well-defined segmented (divisions) of projects, compared to the complexity of IT investments in public sector organisations, which require strong relationships with and interest from various stakeholders. Therefore, SA as applied in modern commercial organisations is often not suitable for a government context and traditional ROI measures will fail in government (Gartner 2004). Even when formal standards and mechanisms are introduced into

---

5   IT spending will grow, for instance, 6.9% year over year to $1.8 trillion in 2012 (Ann and John, 2012)

the processes of SA, success is not guaranteed. The uniqueness of the government sector and its difference to commercial organisations is not only evident in the perspective of the business strategy and the IT strategy, but also in the capture and generation of the public value.

The concerns most frequently cited include fears of a lengthy and complicated implementation, employee reluctance to place information on the system and contribute to its growth, and the difficulty of assessing return on investment (ROI)

(Tremblay 2006, p.1)

It is therefore recommended that to optimise the value of the IT investments, pubic organisations should implement the principles that are part of the SA that recognise and support the uniqueness of the public organisations (Al-Hatmi and Hales 2010; Vogt and Hales 2010).

2.3 DEVELOPMENT OF THEORIES AND RESEARCH ON STRATEGIC ALIGNMENT MODELS

This section introduces strategic alignment models and provides an overview and critique of the key aspects of their potential applicability to public sector organisations.

2.3.1 CRITICAL SUCCESS FACTOR (CSF)

Developed by John Rockart in 1979, Critical Success Factors (CSF) is the first model to link the use of IT to organisational objectives and its strategies. According this method, the following factors are essential: the structure of the industry; the competitive strategy of the organisation; the positioning of the company within the industry; its geographical location; the business environment; and circumstantial factors. Thus, analysing the internal and external environmental factors of the company, including interviewing executives to identify such factors, are the processes that should be used by the organisation in order to achieve competitive advantages. Since the CSF changeover time differs from one organisation to another, this process should be repeated periodically. Taking measurements continuously increases the effectiveness of the CSF method and, as a result of these continuous measurements, the method gained popularity and was used as the strategic planning technique to identify strengths and weaknesses (Sullivan 1985; Pollalis 1993). Rockart (1979) also defined the information needs of CEOs, identifying critical areas of concern such as decision rights and hierarchy responsibilities of IT budget and investments. His definitions, however, do not provide the detailed requirements specification which are necessary in today's organisations. Hence, the CSF method was used to provide a focal point for directing computer-based information system development efforts that pinpoint the key areas that require a manager's attention and thus communicating the role of information technology to senior management. The goal of competitive advantage, however, is less perceived in a public organisation than a commercial one. This is so because the main goal of public organisations is to better service their citizens through technology implementation and to learn from other countries rather than to compete between governments.

Though the Critical Success Factor method has its limitations in measuring IT value and addressing the complexity of transnational information systems in public organisations, it still allows for the identification of areas that can be improved between the functions of information systems and an organisation's objectives. The CSF method is still widely used today to deal with problematic aspects of managerial information systems. However, managerial information systems in this model should be based on an executive's own definitions of the information needs in his or her organisation and should focus only on the limited areas which guarantee an organisation's achievement of competitive performance.

Another issue with the CSF is its suitability when a project becomes complex, especially in a public organisation. The complexity of government projects affects the process of managing, allocating and timing resources to achieve the desired goal in an efficient and expedient manner (Linchko and Calhoun 2003). The objectives that constitute the specified goal may be defined in terms of time, costs, or technical results, goals which cannot be measured using this method because it does not detail requirements specification (Rockart 1979). Additionally, the model becomes difficult to use for organisations whose analysts do not possess the capability to successfully apply the method, which must be directed by skilled analysts. Complex environments, such as those in public organisations, require the upper level of human capacity and ability to effectively share knowledge, which means this model is even ineffective with lower-level managers as the sole means of eliciting information requirements.

Thus, CSF can be considered an earlier model that attempts to measure effectiveness of managerial aspects of information systems but has considerable flaws when used in the public sphere.

2.3.2 STRATEGIC GRID METHOD

In 1984, McFarlan developed a model called the 'Strategic Grid Method'. By addressing four 'quadrants': support; factory; transition; and strategy, each of which represents a situation for the company, McFarlan's model explained how IT is related to strategy and business operations in a company.

This method represents the present and future impact of IT application on the business. To evaluate the strategic impact of IT on business, McFarlan (1984) addressed five questions based on Porter's Five Forces model of strategy:

1. Can IT establish entry barriers for market competitors?
2. Can IT influence the change of suppliers as well as alter bargaining power?
3. Can IT change the basis of competition (based on cost, differentiation or focus)?
4. Can IT alter bargaining power in relationships with buyers?
5. Can IT generate new products?

These questions serve as a guide for the new direction of competition either within the organisation or with other companies. As a result, IT is given a major role in competitive markets in obtaining advantages over competitors (Porter and Millar 1985) and gaining essential competencies (Duhan, Levy et al. 2001). In short, IT can provide advantages such as:

- Entry barriers
- Influencing suppliers
- Competitive advantages
- Power in relation with buyers
- Generating new products
- Identifying key competencies
- Positioning the organisation in the network and in the world
- A clearer level of hierarchy

The recent dramatic impact of IT on organisational performance has necessitated identifying the appropriate method for managing this vital organisational resource. This method, therefore, is used to help management understand how IT impacts firms in different ways prior to selecting management tools. This method can be applied to public organisations, particularly when a comparison is needed between the impact of the IT application of existing systems and the strategic impact of applications development portfolio (the future impact) to leverage its strategy planning. In other words, this model is adopted to indicate the strategic application of the information systems in the organisation. Although the Strategic Grid method is highly recognised and is the most quoted conceptual frameworks in IS literature (Kaplan and Norton 2001; Presley 2006), the method does not provide us with valid operational measures which could potentially have great use in empirical research studies (Hartung, Reich et al. 2000). For example, IS activities may represent an area of strategic concern for some organisations, but for others they may be cost-effective and supportive in nature. When the IS function does not have strategic importance, the trust of its application portfolio will affect IT planning. Thus, the emphasis the model places on various aspects of planning should be different for organisations with different IS environments (the four grids suggested by model) and can be inappropriate for some organisations that variety of focus (Raghunathan and Raghunathan 1990).

2.3.3 THE BALANCED SCORECARD

The Balance Scorecard is defined by the Balance Scorecard Institute (BSI 2011, p.1) as:

..a strategic planning and management system that is used extensively in business and industry, government and not profit organisations worldwide to align business activities to the vision and strategy of the organisation, improve internal and external communications, and monitor organisation performance against strategic goals.

The Balanced Scorecard (BSC) was introduced in 1992 by Kaplan and Norton. It was initially used to analyse the financial measures of past performance with the measures of the drivers of future performance in a more comprehensive system of linked measurements (Mingay 2007) than previous systems, which used ad hoc collections of checklists for managers for non-financial measurement. However, the Balanced Scorecard can be hard to implement, for the public sector in particular, because it is primarily a top-down management tool that tends to hamper bottom-up initiatives (Muniz 2009). There is a challenge in accounting for the strong experienced and creative forces from the lower levels of the organisation.

Therefore, based on the characteristics of the public sector, in its present form the Balanced Scorecard is only capable of covering a portion of the needs of the public sector.

There are five important limitations to the original Balanced Scorecard model (Maltz, 2003):

- It fails to adequately highlight the contributions that employees and suppliers make to help the company achieve its objectives;
- It does not identify the role of the community in defining the environment within which the company works;
- It does not identify performance measures to assess stakeholders' contribution;
- It fails to account for the importance of 'motivated employees', particularly critical in the service sector; and
- The distinction between means and ends is not well defined.

As evaluation methods that rely on financial measures are not well suited to newer generations of IT applications (Tang, Han et al. 2004), changes were made to the BSC by Kaplan and Norton to make it more suitable for its new market. These modifications are: considering the IS department as an internal services suppliers rather than external; recognising IS projects are commonly carried out for the benefit of both end-users and the organisation as a whole (rather than individual customers within a large market); and having performance perspectives targeted at the needs of non-profit organisation (Kaplan and Norton 2001).

Kaplan and Norton's model applies a cause-and-effect measurement to strategy as well as noting the linkage of measurement to strategy. Due to this, the modified BSC is increasingly used in manufacturing and service companies, non profit organisations and government entities. The implementation processes of this model therefore focuses on the following four perspectives (Lankhorst 2004):

- *Customer perspectives* (value-adding view);
- *Internal business process perspectives* by satisfying shareholders and customers by promoting efficiency and effectiveness in our business processes;
- *Financial perspectives* (shareholders' view) by reducing risk and delivering value to the shareholders; and
- *Future readiness and growth perspectives* by sustaining the organisation's innovation and change capabilities through continuous improvement and preparation for future challenges.

In other words, the model focuses on customer satisfaction, internal business processes and the ability to learn and grow and can be used to measure, evaluate and guide activities that take place in specific functional areas of a business. It can also be used to shed greater light on performance at an individual project level.

Yee-Chin (2005) reports that the likelihood of successful BSC implementations in the public sector increases if the target organisation already has a clear vision and strategy, indicating that the BSC is not an effective tool in terms of its strategy-developing aspects. However, this lack in strategy development can be addressed to include different stakeholders in the public sector by complementing the strategy development process with elements from stakeholder theory and incorporating these elements into the Balanced Scorecard.

2.3.4 STRATEGIC ALIGNMENT MODEL (SAM)

Unsurprisingly, many recent alignment models are based on the Strategic Alignment Model (SAM) which was proposed by Henderson and Venkatraman in 1993. This is because the main parameters of this model (the internal and external factors of the company) are still the focus of many organisations today. According to this model, four factors should be considered: business strategy, IT strategy, organisational infrastructure and IT infrastructure. Each of these domains contains an interrelated set of decisions. For example, business strategies include decisions about business scope (product and market offerings), competencies and business governance (a structural mechanisms to organise the business). According to this model, business/IT strategy is considered as an external factor because it relates to business/IT scope, governance and distinctive competence, whereas business/IT infrastructure is considered an internal factor and relates to administrative infrastructure and the architect's processes. Henderson and Venkatraman (1993), and Luftman, Lewis et al. (1993) also identify four perspectives of strategic alignment: executing the strategy, technological transformation, competitive potential and services.

SAM is conceptualised in terms of two fundamental characteristics of strategic management: *strategic fit* - the interrelationship between external and internal domains and *functional integration* – the integration between organisational and technology domains(Laudon and Laudon 2004). Effective alignment requires both strategic and functional alignment. Therefore, SAM is a business/IT management framework and was built to enable the successful implementation of business and information systems and their corresponding infrastructure components (Henderson and Venkatraman 1993; Luftman, Lewis et al. 1993). The model is used to support and shape business policy, elevating IT strategy from its traditional support role as an internal support mechanism (Laudon and Laudon 2004). This model was also used as the basis for Presley's decision-making model (2006): a decisional model developed to assist in the analysis of enterprise resource planning (ERP) systems, which can be used to operationalise the SAM for investment analysis.

According to Henderson and Venkatraman's model, the horizontal dimension of strategy and operations has a direct mutual influence on business and IT. However, aligning business with IT is also influenced by many other factors such as cultural, political, financial and semantic/social aspects. The main limitation of this model is the failure to recognise these aspects. There is a need for a deeper understanding of the aspects in the alignment of the business and the information systems to promote continuous improvement. Thus, in reviewing the SAM, the model has several major gaps. Firstly, there is a lack of assurance and feedback mechanisms in determining the alignment at the strategic level in the business/IT operational settings. Secondly, there is an uncertainty in measuring the defined strategic alignment that explains the impact on these strategic perspectives of current organisational practices within the operational levels. When given a specific alignment case study, there are no concrete criteria to determine which of the alignment perspective(s) is most important in the case (van Eck, Blanken et al. 2004). Moreover, the SAM is not a constructive theory of strategic alignment; it does not provide any guidelines on how to reach specific alignment goals. Finally, this model fails to identify the elements or perspectives in the alignment model that involve the execution of strategy, the potential of the technology, competitiveness and the service level (Gutierrez, Orozco et al.

2006). Therefore, further investigation is necessary in order to prove that the alignment is a factor in the business/IT strategy.

The above gaps can be minimised by combining the SAM into the range of strategic choices to assess the capabilities of managers to face the current changes, interrelate into situation and to explore the other potentials of the workplace (Ward and Peppard 2002). Due to the discussed limitations, some studies have extended and refined the original model. For example, Maes (1999) developed two extensions to SAM, adding a layer between strategy and operation (termed structure) and a column between business and IT (termed information/communication) (See also Section 2.3.6). Luftman (2000) created another a practical guide extension by either examining current business and IT alignment in a firm or discerning deficiencies in the required alignment.

2.3.5 MANAGEMENT BY MAXIM

In 1997, Broadbent and Weil developed another proposal, similar to Rockart's CSF model, called 'Management by Maxim'. Management by Maxim consists of business maxims, IT maxims and IT infrastructure and strategic context. Whilst the business maxims address what to coordinate across the firm, what to leverage from within business units and what to leave to medium or low management decision options, the IT maxims describe how a firm needs to connect, share and structure information and deploy IT across the firm. A term to express this synbook is 'maxim' (business or IT maxim).

The Management by Maxim model (Broadbent and Weill 1997) is used to help managers identify business and IT maxims that can help them to determine the IT infrastructure capabilities necessary to achieve their business goal. This model also helps to create a business-driven IT infrastructure that involves decisions based on a sound understanding of a firm's strategic context. This understanding can be achieved and communicated through business maxims. According to this model, business maxims capture the essence of a firm's future direction and eventually lead to the identification of IT maxims that express how a firm should deploy IT resources and gain access to and use information. IT maxims provide a basis for a firm to make decisions on its IT infrastructure services. This IT infrastructure provides the shared foundation of IT capability for building business applications and is usually managed by the Information Systems (IS) group. Since IT infrastructure investments are long-term commitments that can account for more than 58 per cent of the total IT budget of large firms (Broadbent and Weill 1997), decision-making concerning IT infrastructure investments needs to be closely related to its capabilities to underpin the competitive positioning of business applications, such as cross-selling opportunities, marketing new products to customers and improving redesigned cross-functional processes.

To identify gaps in business maxims, the firm must focus on the competency of the employees and coordination of the employees with the firm as well as the resources and the implications of management of information and IT. In identifying the gaps in IT maxims, the firm still needs to connect, share and structure the information and its deployment across the firm. Therefore, there are clear leadership and management challenges in business and IT. This means that there should be proper training for the managers to orient them to the role of IT, related processes and accessibility (Broadbent and Weill 1997). In addition, the board of directors often makes decisions about the business cases for IT infrastructure investments based on technical criteria presented to them rather than in the context of long-term business needs, which are often not clear due to the many options for configuring IT investments. These challenges are difficult for IT managers, and this is particularly evident in public sector organisations where governments decide on long-term investments in terms of enabling infrastructures, with business benefits not based on cost reduction. As such, technological factors are still dominant in influencing infrastructure decisions (Whitten, Bentley et al. 2004; Mingay 2007).

Another obvious challenge is determining and measuring the effective operations gained from the

IT investments (Broadbent and Weill 1996). As the Management by Maxims model is supported by the capabilities in which the limitation may be traced, one challenge that may arise is the fact that each IT investment is supported by the firm's mission and values, which in turn are supported by the current business plan. Hence, identifying and developing the appropriate infrastructure is actually a true challenge in this model.

2.3.6 THE GENERIC FRAMEWORK

The main limitation in Henderson and Venkatraman's model, SAM (Section 2.3.4), is that it states that the horizontal dimension (Strategy and Technology, see Figure 2-1) has a direct mutual influence of business and IT. The process of aligning business to IT, however, is also influenced by many other factors such as cultural, political, financial and semantic/social aspects. To address this limitation, the Generic Framework Alignment was proposed by Maes in 1999. In this model, two attributes were extended from the Henderson and Venkatraman's model: 'Structure' in the horizontal level and 'Information and Communication' in the vertical level (Figure 2-1). These variables are dependent on each other and are both considered essential to successfully managing and designing modern alignment between business and IT within the organisation. They represent the structural components (variables) related to competencies and infrastructures of the organisation.

FIGURE 2-1: A GENERIC FRAMEWORK ADOPTED FROM HENDERSON AND VENKATRAMAN'S MODEL OF ALIGNMENT

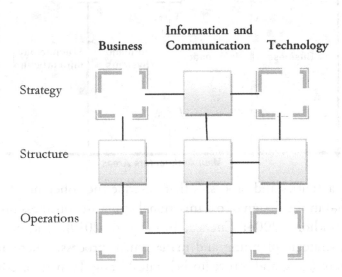

This Generic Framework is used to provide a clear direction of the perspectives to be considered at the different levels of redefining business/IT alignment through unified framework components (Maes, Rijsenbrij et al. 2000).In Maes' framework model (1999), the strategy level deals with decisions regarding scope, core capabilities and governance. The structure level is basically concerned with architecture and capabilities and the operational level with processes and skills. Different roles, therefore, can be derived at each level while the business/IT relationship remains key to information management at different levels of this model.

In assessing the Generic Framework and measuring its effectiveness in investigating the strategic alignment of business/IT strategy, there are difficulties that might be experienced in applying this model in a public organisation. The Generic Framework is more relevant for examining external factors. Therefore, if the generic framework is utilised with the business/IT strategy in the public sectors, some resources (the internal resources) might not be addressed holistically. The government needs the idea of accountability in order to promote its transparency; otherwise accountability can be difficult to address. Additionally, the uncertainty in the sources of information makes it more difficult to improve the resource management and the management of risk related to IT, explaining the result magnitude of loss or gain of an organisation(NASCIO 2007; Stewart 2008).

2.3.7 THE INTEGRATED ARCHITECTURE FRAMEWORK (IAF)

Architectural design plays a significant role in business/IT alignment. The Integrated Architecture Framework (IAF) was designed by the Cap Gemini Institute (2000), to support the role of IT in existing business processes. The model consists of three dimensions. The horizontal dimension concerns the four main architecture areas: business processes, information, information systems and technology infrastructure, while the vertical dimension concerns the five design phases supported by the architectural description (see Figure 2-2). The third dimension is called 'special architectural viewpoints' and it deals with all architecture areas.

FIGURE 2-2: THE INTEGRATED ARCHITECTURE FRAMEWORK (IAF)(MAES, RIJSENBRIJ ET AL. 2000)

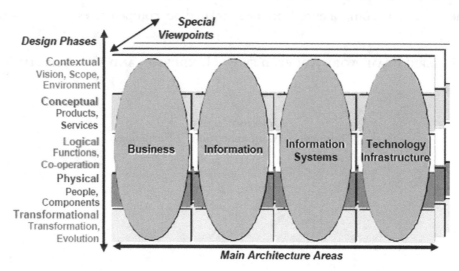

This model is used to define activities and processes that describe the coherent whole of the principles, methods, and models that are used in the design of organisational structure, business processes, information systems, and infrastructure (Lankhorst 2004; Cuenca, Boza et al. 2010). The outcome of Enterprise Architecture (EA) is evolving a strategic planning and management process, where an EA framework is applied to describe both the current (as-is) and future (to-be) states (Tang, Han et al. 2004). The framework should also help to articulate how the different perspectives of the architecture relate to one another (Martin and Robertson 2002; Bittler and Kreisman 2005) and facilitate modelling views of enterprise complexity to be humanly comprehensible in its entirety (Martin and Robertson 2002).

The implication of change in management is that change usually occurs in a top-down hierarchy that deliberately affects both the people and the processes of a business. The problem with this model is that adoption of government IS/IT is not straightforward and cannot be done in a short period of time; rather it requires an integrative architecture approach to be in place for government information and online services adoption. The government IT adoption also requires significant changes in organisational infrastructure, which can engender resistance in government IS/IT adoption (Ebrahim and Irani 2005). This is one of the reasons public sector organisations are required to integrate the high-level vision and structure of the business as well as the system applied in supporting the alignment.

Additionally, public organisations have a tradition of processing information systems in separate databases, which means that the transactions and operations of each department are not connected or linked to each other. At the same time, there are different IT management and related decision levels involved (for example, local and central level, upper and medium level). This creates barriers between an organisation's systems and processes concerning information transmission and communication and the integration of government in terms of the database systems (Ebrahim and Irani 2005).

In order to bridge the gap in the developing environment of organisations, there should be a clear and holistic vision of how systems can aid organisations works towards IT success (Macaulay 2004).

2.3.8 THE SOCIAL DIMENSION OF ALIGNMENT MODEL

Alignment is the degree to which the information technology mission, objectives, and plans support and are supported by the business mission, objectives, and plans(Reich and Benbasat 1996). According to this definition, alignment is a state or an outcome. Alignment is, however, likely to be a dynamic process expressed through communication and planning. Previous works focus on the relationship between alignment and IT performance (Chan, Huff et al. 1997; Subramani, Henderson et al. 1999) or between shared knowledge and IT performance (Nelson and Cooprider 1996). In 2000, Reich and Benbasat proposed a new model that focuses on the factors that create or inhibit alignment, called the Social Dimension of Alignment model. Their study is supported by an interpretive approach by Klein and Myers (Klein and Myers 1999),who investigated how certain critical factors interact to create conditions that enable or inhibit alignment.

Reich and Benbasat (2000, p.81) defined the Social Dimension of Alignment as "the state in which business and IT executives understand and are committed to the business and IT mission, objectives and plans". They also identified the main four factors that influence the social dimension of alignment as follows:

- Shared domain knowledge between business and IT executives
- IT implementation success
- Communication between business and IT executives
- Connections between business and IT planning processes

All these perspectives were found to influence short-term alignment except shared domain knowledge, which was found to influence long-term alignment. It was also found that a new strategic business plans perspective influenced both short- and long-term alignment.

The significant role of the social dimension of alignment was recognised earlier in previous literature: namely the social construction of reality (Berger and Luckmann 1967); the typology of seven techniques (Galbraith and Schendel 1983); a rational model of organisational decision-making; the political behaviour model and resource dependency model (Boynton and Zmud 1987); the strategic vision and strategic learning approaches that integrate a number of actors (Mintzberg 1993); the social dimension of alignment (Nelson and Cooprider 1996; Subramani, Henderson et al. 1999); and 'cultural gap' (Taylor-cummings 1998).The 'culture gap' between IT and business people, as identified by Reich and Benbasat(2000), has been identified as a major cause of system development failures(Reich and Benbasat 2000).

In defining the relationship between an organisation and IT functions, IT managers have identified the existence of a culture gap between the groups of communicators(Reich and Benbasat 2000). IT managers believe that it is easier for IT staff to understand the business than it is for business staff to understand the problems of their IT counterparts in order to enhance a short-term alignment. It has been shown that the non-IT staff often misunderstand the role of the IT staff, as it is sometimes difficult for those outside IT to perceive job function(Riege and Lindsay 2006). From this perspective, the issue that can be the most problematic is handling staff, even though the managers clearly know the differences between the two groups (Khosrow-Pour 2006).In IT/business units without a high level of shared domain knowledge, failed or failing IT projects result in finger-pointing, reduced levels of communication and low levels of short-term alignment(Reich and Benbasat 2000).

Thus, a prerequisite for higher alignment is that IT personnel and business staff collaborate at all levels of an organisation. The difficulties identified with this are the invisibility of IT staff, barriers in communication, the history of IT and business relationships, the attitudes of the members of the organisation to IT, the domain of knowledge and the application of leadership strategy.

It is important at the onset of collaboration to acknowledge and address potential obstacles and pitfalls that may be presented along the road to collaboration.

(NASCIO 2007, p.2)

Additionally, there is great diversity in the cognitive structure and content of a business that can result in inconsistency in the level of alignment, thus reducing the success of IS planning (Chan and Reich 2007).

2.3.9 THE SAMM

Luftman (2000) argues that strategic alignment refers to applying IT in an appropriate and timely way in harmony with business strategies, goals and needs. His definition differs from that of Reich and Benbasat (2000), which focuses on IT plans and social aspects rather than the application of IT (Shimizu, de Carvalho et al. 2005, see also Luftman 2000).

Luftman (2000) developed a Strategic Alignment Maturity Model (SAMM), with 12 components to be taken into consideration in terms of the defining the business/IT strategic alignment. As Figure 2-3 indicates, the 12 components fall under four dimensions as follow:

- *Business Strategy:* related to market and products
- *Organisational Infrastructure and Processes:* related to the employees
- *IT Strategy:* related to how the authority provided the resources as well as the responsibility
- *Information Technology Infrastructure and Processes:* related to the IT skills and knowledge of the people

FIGURE 2-3: LUFTMAN'S TWELVE COMPONENTS OF ALIGNMENT (2000)

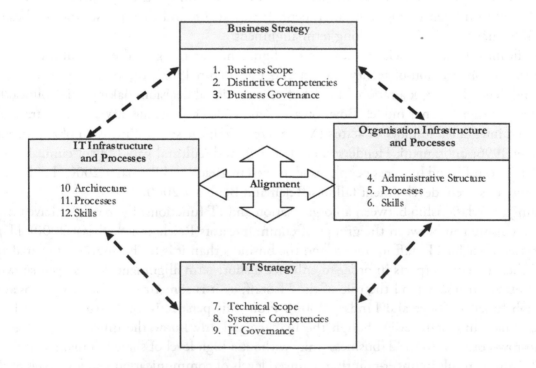

Luftman provided a further six perspectives for evaluating business/IT alignment, which he described as 'alignment components'. The presence of poor performance in these perspectives inhibits strategic alignment. These six perspectives are:

- Communication
- Competency/value measurement
- Governance
- Partnership
- Scope and architecture
- Skills

Unlike other models, this model combines the descriptive and prescriptive aspects of alignment. This unique combination generates a roadmap that practitioners and consultants can follow to attain higher levels of IT effectiveness, which in turn can lead to better business performance(Council 2007). The model can be used as an instrument to better leverage IT, and the application of this model has been used in both profit and non-profit organisations to help management to understand how to improve their current practices (Gutierrez, Orozco et al. 2006).

2.3.10 KNOWLEDGE-BASED THEORY OF ALIGNMENT

The Knowledge-based Theory of Alignment was developed by Kearns and Sabherwal (2007) to understand the effect of context and planning behaviours on business/IT strategic alignment and the effect of high-level managers' knowledge of IT on strategic alignment by supporting the sharing of domain knowledge. They theoretically and empirically examined the following factors in relation to strategic alignment as a whole in an organisation:

- *Planning behaviour* (IT managers' participation in business planning and business managers' participation in IT planning)
- *Upper level management knowledge of IT*
- *Contextual factors* (organisational emphasis on knowledge management and centralisation of IT decisions in regard to business/-IT strategic alignment)
- *IT projects* (quality of IT planning and implementation problems in IT projects)

A study conducted by Hartung, Reich et al. (2000) revealed that even though the model was developed for the for-profit industry, it can equally be applied to public sector organisations. The study results confirmed that the shared domain of knowledge and communication was positively related to alignment. The role of leadership also proved to be important.

Applying this model to public organisations, however, is not an easy task. Due to the diversity of stakeholders, there is always a shortage of IT skills and human resources and insufficient training offered by the organisation. Other limitations found by Hartung, Reich et al. (2000) in their research into a Canadian military organisation include:

- The organisation had only recently begun to engage in long-term strategic planning and thus had a lack of maturity in their planning processes;
- Where there was misalignment, people could not articulate how IT would improve the situation; and
- The concept of 'shared domain of knowledge' required a high degree of cross-functional experience and many staff did not have enough experience in their current station to judge the situation.

Overall, in this example, public organisation alignment appears to be a function of shared knowledge and communication, but it also has an impact on short-term alignment that affects staff.

2.4 Gaps in Strategic Alignment Models

Overall, the concept of misalignment between business strategy and IT strategy in the literature proves to be affected by many factors: a cultural gap[6] (Grindly, 1992), reconciling the business/IT relations (Ward and Peppard, 1999), measuring business/IT alignment (Luftman, 2003), shared domain knowledge (Hartung, Reich et al. 2000) and leadership accountability (ITGI 2003; Guldentops 2004). The literature has shown that one or more of these factors can play a role in creating the misalignment between the business strategy and IT strategy (Shamekh, 2008), depending on context. These factors (referred to as perspectives and attributes in this book) include:

- The availability of the Management of Information Systems (MIS) plan tool to measure Critical Success Factors (CSFs) (a[7]);
- Measuring performance, establishing corporate activities (milestones) (b);
- Current and future strategic impact of application of IT (c);
- Competitive advantages (cost, establish barriers, suppliers, et cetera) (d);
- Customer perspectives (value-adding view) (e);
- Internal perspectives (process-based view):satisfying shareholders and customers by promoting efficiency and effectiveness in our business processes (f);
- Financial perspectives (shareholders' view):delivering value to our shareholders (g);parenthetic alphas
- Optimisation of supply chain (h);
- Learning and growth perspectives (future view): sustaining our innovation and change capabilities through continuous improvement and preparation for future challenges (i);
- Strategic fit: the linkage between internal and external domains (j);
- Functional integration: the integration between business and IT domains (k);
- Business/IT strategy (l);
- Organisational/IT infrastructure (m);
- Information and communication, decision-making (n);
- Value management, governance, technological and organisational capabilities (long-term strategic alignment) (o);
- Business/IT maxims (IT resources, IT infrastructure capabilities for building business applications, competitive positioning, IT infrastructure decisions based on an organisation's strategic context) (p);
- Scope, core capabilities and governance (q);
- Information and communication: shared domain knowledge, top managers' knowledge, communication between IT-business managers and executives communication, communicating the role of IT to senior management (r);
- Enterprise architecture (organisational structure, business processes, modelling processes of information systems) (s);
- IT implementation success, value measurement (t);

---

6   A culture gap is an attribute that explains the problems that can exist between the IS/business function. Ward and Peppard (1999) considered it as a convenient description of the symptoms but not an explicit cause of the gap.

7   Alphabets serves to synthesize low or high attributes according to each SA model in public context given in Table 2-2.

- Quality of IT/business plan (quality of IT planning and implementation problems in IT projects) (u); and
- Skills, and planning behaviour (IT managers' participation in business planning and business managers' participation in IT planning) (w).

Section 2.5 discusses the major gaps in the models outlined in Section 2.3, with reference to the list above, examining the implications of each model's limitations when applied to public organisations. In this regard, gaps related to models, engaging stakeholders, effective IT governance and the public value of IT investment with reference to public sector organisations will be highlighted.

2.5 GAPS OF STRATEGIC ALIGNMENT MODELS IN PUBLIC ORGANISATION

It is a known fact that important and costly investment resources in Information Technologies and Communications (ITC) are wasted and badly designed, with functions not needed by public enterprises.   (Muniz 2009, p.21)

Information systems (IS) are defined as a set of components (people, process and data/relational mechanisms) that retrieve, store, process and distribute information to support the decision-making process and IT strategic planning (Laudon and Laudon 2004). These systems may incorporate information technology among their components (Whitten, Bentley et al. 2004). Governments' management of IT projects, therefore, face numerous challenges, for example excessive cost, project failure or unrealised value (Siau & Long, 2004). Most of these challenges occur due to the failure of these models to account for the other forms of public value (Shin 2002; Davis, Dehning et al. 2003; Gartner 2007) needed for cross-sharing knowledge and engaging diversity of stakeholders in the ICT decision-making process (Brown and Grant 2005; Elpez and Fink 2006; NASCIO 2008) and for the mechanisms necessary to establish effective IT governance in public organisations (Chen, Chen et al. 2007).

The insufficient IT security compliance associated with IT project failures also calls for more transparency in public regulations, which requires a cooperative approach among the state[8] CIO, executive management and human resources training to address the implications for state IT threats from employee behaviour that is inconsistent with organisational ICT policies (NASCIO 2007).

In this section, the concept of misalignment has been identified and discussed. Next, the four dimensions of these models are grouped according to their incapability/misalignment in terms of the potential application of these models to public sector organisations as identified in literature. These dimensions, addressed within the context of the public organisation, are: model limitations, social engagement, IT governance and application for service quality instrument to IS (public value).

2.5.1 GAPS RELATED TO MODELS

The theory of Strategic Information Systems Planning (SISP) considers alignment as one of the key rational success dimensions (Lederer and Salmela 1996; Segers and Grover 1999). Reich and Benbasat (2000) further define the major factors that impact on social alignment as integration between business planning and IS planning, shared domain knowledge, communication and past IT implementation success. The roles played by IT executives are also cited as an important issue (Kajalo, Rajala et al. 2007). Luftman (2000) identifies the enablers of IT implementation adoption including senior executive support of IT, IT involvement in strategic development, IT understanding of the business, effective business/IT partnership, well-prioritised

---

8   NASCIO's literature refers to a government organisation as a state

IT projects and IT that demonstrates effective leadership. However, based on Luftman's study (2011)which appears as the most pertinent study concerning strategic alignment, the alignment maturity of business and IT is the major impediment to SA.

Cohen (2003) combined previous findings about the intellectual and social dimensions of alignment into four integrated factors that affect alignment:

- Business/IS planning
- Rational-adaptation in SISP
- IT implementation success
- IT managerial resources

Other earlier studies by Broadbent and Weill (1993) and Earl (1993) suggest that strategic alignment has a direct effect on performance. Further studies conducted by Palmer and Markus (2000) and Sabherwal and Chan (2001) found that it was only under certain contingencies, such effective communication and leadership, that SA leads to better performance. The cumulative evidence of the SA research, however, points to a positive linkage between SA and performance (Subramani, Henderson et al. 1999; Chan and Reich 2007) although there are no predetermined concrete criteria to determine which SA perspectives contribute to an organisation's performance (Bhansali 2007). Critics of these earlier models point out that strategic alignment is not a static point; rather alignment is now considered as a process, since both IT and business strategy are facing a dynamic environment(Boynton and Zmud 1987; Eisenhardt and Schoonhoven 1996).

Many of the strategic alignment models, such as The Balanced Scorecard or Management by Maxim, are derived from empirical studies conducted in the business sector. For example, Broadbent and Weill (1993) studied the Australia banking industry; Reich and Benbasat (2000) studied 10 Canadian insurance companies; and Kearns and Sabherwal (2007) studied more 1,100 companies in the US. Since the 1990s, these models have shifted their focus from resources, automation, marketing, processes, knowledge, IT investments and value to the more holistic functional applications of a whole organisational context. As discussed previously in Section 2.3, the IT architectures in these models operate with different set of priorities and values that are not suited to a contemporary government context. They do not readily reflect the enterprise-wide integration and collaboration that is necessary for government models (Brown and Grant 2005; Weerakkody, Janssen et al. 2007).

2.5.2 Gaps Related to Engaging Stakeholders

Stakeholder and constituent engagement is important in improving public services. NASCIO (2007) presented a strong recommendation for cross-boundary collaboration and information-sharing in order to acknowledge and address potential obstacles and pitfalls that may occur in government IT initiatives. Past research also identified the end-users who will ultimately use the system and the IT professionals who deliver those systems as strongly contributing to project success. Fisher (2001, p.25) pointed out "If users cannot use the system effectively and efficiently it cannot be deemed to be a success". Hannaway and Hunt (1999) and Carr and Folliard et al. (1999) also argued that programs (projects) implemented in isolation without effective communication with key stakeholders would result in failure.

Similarly, the case study of the National IT Literacy Program (NITLP) in Singapore undertaken by Tan and Pan (2003) showed the importance of the organisation-stakeholder relationship for the successful implementation of eGovernment initiatives. Systematic collaboration and high-level accountability amongst executives and other stakeholders is now considered to be at the core of success of any IT project. Soeparman, Duivenboden et al (2009, p.246) define the concept of collaborative innovation as:

> The enduring and collective ability of organisations to cope with and anticipate changing circumstances in their environment and/or to improve performance by renewing or altering the way core-activities, such as service delivery, are interweaved and joined-up.

Thus, public service delivery is often characterised by the need for collaboration amongst all stakeholders. A study by Fardal (2007) also supports the vital role of collaboration between ICT strategy and ICT users to identify factors to improve ICT project processes. And as Maylor, Brady et al. (2006, p.663) noted, "Nowadays, it is hard to imagine an organisation that is not engaged in some kind of project activity". Evans (2004, p.303)also notes that failed IT projects are common in organisations, as IT solutions often do not meet the expectations of business clients. She goes further to say:

The relationship between business and IT suffers because IT projects fail to solve the real business need and, on the other hand, many IT projects fail because of existing interpersonal relationship problems between business and IT employees.

As mentioned by a participant of Evan's study, there is a problem of work overload and insufficient training.

Some of the technology is too complex for business managers to understand entirely. Insufficient resources can lead to project failure. One problem is that the permanent staff are not valued as they should be. Some business people don't understand the full potential of the IT capabilities in the sense of what is possible and what is not.

It is also believed that participation in the budget-setting process, its target and its acceptance by budget holders improves the attitude of middle managers and hence is crucial to the success or failure of a budgeting control process (Dodd, Yu et al. 2009).

Another gap identified in strategic alignment in the public sector is the insufficient skills of human resources staff. Getting the right people with the right skills at the right time is a major issue in large projects. Frameworks such as the Integrated Architecture Framework (IAF) have proven to be a great help in achieving human resource-related objectives but, because of the rapid change in the technological environment and the low pool of qualified staff applying for positions, the public sector experiences difficulties in acquiring appropriate staff.

2.5.3 GAPS RELATED TO EFFECTIVE IT GOVERNANCE

The objective of IT governance in government is ensuring that the government is effectively using information technology in all lines of business and leveraging capabilities across boundaries appropriately to not only avoid unnecessary or redundant investments, but also to enhance appropriate cross-boundary interoperability(NASCIO 2008). As a result of a number of major corporate and accounting scandals that cost investors billions of dollars, including those affecting Enron, Tyco International, Adelphia, Peregrine Systems and WorldCom, many governments developed strict regulations or applied existing or developing frameworks. These were introduced and enforced to ensure that companies provide greater transparency and accountability, and include the Sarbanes-Oxley Act in the United States, AS-8015 in Australia or ITIL in Britain. A consequence of this is that there is now more clarity of direction and consistency in setting goals and measures of success are consistent with the public interest.

According to several IT Governance Institute Reports (ITGI 2003; ITGI 2008; ITGI 2008), IT governance implementation overall in the private sector is higher than in the public sector. The reason for this difference is perhaps due to fact that the financial values realised from IT by financial sectors occurred earlier in private companies than in the public sector. The non-financial value realisation benefit from IT governance occurred later in government, with its limits in public service spending. While according to the IT governance Global Status Report(ITGI 2008) the importance of IT governance in all sectors continues to increase, the best practice frameworks of IT governance in governments have not been fully investigated (Chen, Chen et al. 2007). The scarcity of other studies undertaken in Australian organisations indicate that little research has been conducted into examining what mechanisms contributed to establish effective IT governance within public sector organisations(Green and Ali 2007).

Effective leadership (IT governance structure) is vital to the success of any project and the absence of top leadership support is often a key factor for business performance failure (Allen and Kraft 1987; Brown 1995).

Organisational leaders, however, tend to be the most resistant to change (Dyer 1986; Brown 1995). They have enjoyed success in the past and they are convinced that any new systems will be a threat to their positions. Kane (2005) argues leadership is the most critical ingredient in any change effort, yet many employees do not consider their top executives to be effective. The aspect of leadership (organisational structures) investigated in this study will be described in greater detail later in Section 4.6.

Looking at all of the issues related to organisational aspects, it is easy to see how misalignment can occur (Wieringa, van Eck et al. 2004). However, there are other issues related to management aspects in public organisations that are harder to manage, such as the performance of the employees or staff, the culture of the people and unstable business goals or objectives(Khaiata and Zualkernan 2009).

According to De Lone and McLean' model (2002), which was later adapted by Beynon-Davies (2002), the success or failure of government ICT systems is related to the functionality, usability and utility of IT. De Lone and McLean (2002; 2003), however, acknowledged that defining IT project success is difficult as different researchers focus on different aspects of success, making comparisons difficult and the prospect of building a cumulative view for IS research similarly elusive.

Furthermore, Elpez and Fink (2006) argue that the success of an IS project in a public organisation is whatever the stakeholders perceive it to be. This means that different stakeholders can view the project outcome from various perspectives and arrive at different conclusions, which in turn provides more challenges in reaching an acceptable definition of IS project success.

Many projects suffer when project participants are in disagreement as to the proper success emphasis or goals for the project. These differences in success emphasis are a result of poor team alignment.

(Griffith and Gibson Jr. 2001, p.69)

Gichoya (2005) divided the factors affecting the successful implementation of ICT projects in government into two main streams: enablers or inhibitors. Enablers are those occurrences whose presence or absence determines the success of an ICT project. The factors that when present reinforce the successful implementation of IT projects include:

- Vision and strategy
- Government support
- External pressure and donor support[9]
- Rising consumer expectations
- Technological change, modernisation and globalisation
- Effective project, coordination and change management
- Good practice

Inhibitors[10] are those occurrences that constrain the smooth implementation of ICT projects in government. The factors that hinder ICT implementation include:

- Infrastructure
- Finance
- Poor data systems and lack of compatibility
- Skilled personnel
- Leadership styles, culture and bureaucracy
- Attitudes
- User needs
- Technology
- Coordination

---

9   The study occurred in an east African country where donation plays a big role.

10   Some inhibitors do not necessarily prevent the implementation of ICT projects but they do prevent advancement and restrict successful implementation and sustainability.

- ICT Policy
- Transfer of ICT idolisers
- Donor push

Both enablers and inhibitors have been confirmed by literature (Saul and Zulu 1994; Luftman 2000; Gartlan and Shanks 2007). But enablers and inhibitors need to be tightly linked to IT and business priorities which lead to shared ownership and shared governance of IT projects. As literature indicates (Shpilberg, Berez, et al 2007) inefficiencies in IT will not corrected by alignment alone. In practical term, a combination of effectiveness and alignment enables agility and improved competitiveness.

The pre-project planning effort involves alignment of the project team with the business needs to develop an adequate scope definition. Ignoring this phase can lead to project scope changes, cost overruns and longer schedules. Therefore, the greater the pre-project planning effort, the greater the chance for project success (Gibson and Hamilton 1994). In addition, the study titled 'Successful Program Management in the U.S. Federal Government' identified four major success factors in successful programs: active high-level executive support; a culture of communication; stakeholder engagement; and agility (NewsRx 2011).

Given that strategic information systems alignment is viewed as essential to organisational success in deriving value from IT investments, research shows that achieving alignment is difficult to attain(Grant 2003). Guldentop (2011) acknowledged that alignment is best achieved when cross-functional, collaborative information systems are instituted.

2.5.4 Gaps Related to Complexity of Value of Government IT Investments

It has been proven that sound project management practices significantly improve the success rates of IT implementations.

(Matt Miszewski, NASCIO 2005-06 president2006, p.1)

A government is a complex public organisation where the effective management of IT projects faces numerous challenges, such as excessive cost, project failure and unrealised value (Siau and Long 2004). Public sector organisations often have more diverse stakeholders than commercial organisations with different social, economic and political priorities. In addition, many public sector organisations face extra challenges including a legacy of rigid bureaucracy, established illogical time-consuming routine tasks, a lack of harmonising integration and poor coordination of different information systems(Weerakkody, Janssen et al. 2007). These problems are common in IT implementation, particularly in many developing countries (Siau and Long 2004; Chen, Chen et al. 2007).

Despite the difficulty of evaluating ICT projects, governments are finding that the inability to demonstrate a public value proposition is becoming politically unacceptable(Guldentops 2004; Crawford 2009). The scope and scale of government investment in IT and the associated IT projects' problems certainly deserve attention. Examples of government expenditure in IT as provided by Cresswell (2006) shown in Table 2-1.

Table 2-1: Government IT Spending

Year	Government	Current Spending	Planned increase in spending
2006	Asia-Pacific region governments (excluding Japan)	U.S.$22.7 billion	U.S.$31.7 billion in 2010

2005	European governments	U.S.$110 billion (U.S.$26 billion in UK alone, approximately 40% above German and France	U.S.$119 billion in 2007
FY 2004	U.S.	U.S.$55 billion for IT investments	U.S.$62.4 billion in FY 2009
2004	China	U.S.$5 billion	Increase 16% in 2009
2002	India	U.S.$943 million on eGovernment	Increase 15% a year to U.S.$3.3 billion by 2009

These examples indicate that the level of IT investment in governments worldwide is increasing on a very large scale. The high expenditure on technology and the increasing failure of IT projects in many countries clearly identifies the need for a more proactive involvement in the control of IT activities. The reason for expenditure scrutiny is explained by Andrea Di Maio, a Gartner Vice President, focusing on the public sector:

If governments do not accurately measure the full value of their IT investments; they risk a serious political backlash. They will be accused of wasting billions of pounds of taxpayers' money on unnecessary technology.

(De Souza, Nariwawa et al. 2003, p.17)

Private organisations also face problems implementing their IT projects. Ryan and Harrison (2000) noted that that over 50 per cent of IT implementations actually cost more than twice their original estimate. When such problems occur in public organisations, it is usually due to the lack of foresight in the IT investment decision process (Schniederjans & Hamaker, 2003) or due to human and organisational issues (Gartner, 2004; Gutierrez, Orozco, Serrano, & Serrano, 2006). A lack of control mechanisms for IT activities is also dangerous when considering the success of an IT project (Nolan & McFarlan, 2005). As a result, many IT projects still fail. For example, Enron in the U.S. lost U.S.$80 billion in 2001 due to software failure (Gelinas, Sutton, & Fedorowicz, 2004) and Nike lost U.S.$200 million due to the failure of the process of implementing supply chain software (ITIL 15/10/2007).

Since government organisations have faced various difficulties in their IT projects, managing the processes of budget, time, and scope is important to determine the extent of business/IT strategic alignment among government sectors. The success or failure of such processes will determine the accomplishment and realisation of public value. The public benefits by which success can be measured are ROI, efficiency, effectiveness and the capability of public organisations to provide the types of IT customer services needed by the community, mostly often referred to customer service, value realisation or public value (Gartner 2004; Institute 2006; ITGI 2008; Luftman 2009; Al-Hatmi and Hales 2010; Al-Hatmi and Hales 2011). As a service, IT supports and enables the organisation in achieving its strategy, values and objectives. Thus, strategic alignment is a process by which the organisation realises certain benefits or values. Chang, Hsiao et al's study (2009) on service-oriented organisations found that SA leads to better customer service quality and is more significant in improving customer service quality than operational and social alignment.

Private organisations are also concerned with values but there are inherent differences that suggest that the same approach to strategic alignment does not apply in both the private and public sectors. For example, the time allowed for implementation and the financial capacity of the public sector is often questioned. If the challenges that an organisation meets in the middle of the IT implementation and business collaboration are considerable or difficult to solve, the postponement of the project is highly possible. Gershon (2009), as quoted by Lundy (2009), points out that public organisations should realistically evaluate the benefits that they might earn from their potential projects. With the application of strategic alignment perspectives and enhanced IT/business strategy relationships, an organisation can maximise the benefits that they expect from IT.

Governments do use various financial-based parameters to assess whether IT investments can deliver the expected value. Such parameters include payback periods, Net Present Value (NPV), Internal Rate of Return (IRR), Economic Value Added (EVA) and Return on Investment (ROI). However, the study by Shin (2002) on the impact of information technology on financial performance indicates that investment in IT does not in itself ensure profitability for an organisation in financial terms. Due to the nature of the relationship between governments and their various stakeholders, these financial metrics are less effective or less valued in a public environment as they do not account for other forms of return based on political issues, effectiveness, efficiency of government services and the achievement of social objectives; collectively called 'public values'. Traditional ROI measures do not apply in government (Gartner, 2004). This problem has been observed other different studies, for example Morgan and Schiemann (1999), Tan and Pan (2003) and Kaplan and Norton (2004) and Firth, Mellor et al. (2008). Other findings by Davis et al. (2003) indicate that payoffs from investments in information technology are difficult to recognise. Therefore, relating IT initiatives to public values remains a challenge to IT and non-IT executives alike.

In response to the limited effectiveness of traditional measures of success in a public sector environment, governments have started to develop their own public value frameworks that attempt to measure the public value of IT investments, such as the Demand and Value Assessment in Australia (AG 2004), the Federal EA Performance Reference Model in America (FEA 2007), WiBe in Germany (Government 2009), MAREVA in France(Bhatnagar and Singh 2010) and eGovernance Assessment Framework (EAF version 2) in India(Rao, Rao et al. 2004). Gartner(2007) defines the public value of IT (PVIT) in terms of three dimensions:

- **Constituent service level**: this level measures the impact on time and cost to users or beneficiaries of government services in terms of accessibility, quality, and convenience of services;
- **Operational service level**: this level measures the internal impact of an initiative both on individual departments and across government organisations; and
- **Political return**: measures the impact of investments on political goals and consensus, the overall impact on society in terms of wider reach of information, alleviation of digital and cultural divides and the impact of economy.

Cresswell et al (2006) undertook five case studies that examined how government investment projects came to deliver value to the public. The projects were the Integrated Enterprise System in the Commonwealth of Pennsylvania and the Washington State Digital Archives in the U.S., the Merkava Project in Israel, the Australian Federal Budgeting and Bookkeeping System in Australia and the IT Service New Brunswick Project in Canada. They found that government IT investment generates public value in two ways:

- By improving the value of the government itself from the perspective of its citizens; and
- By delivering specific benefits directly to persons, groups, or the public at large.

Based on the discussion above, government IT projects face many challenges, including difficulty in measuring the intangible public value of IT investment, engaging a diverse range of stakeholders, leadership and various critical success factors identified in alignment literature. Moreover, projects those concentrate of technology and focus on project's timely completion, and within budget specified, rather than end-users, are more likely to fail. 'Aligning a poorly performing IT organization to the right business objectives still will not get the objectives accomplished' CIO of Selective Insurance Group as cited by Shpilberg, Berez et all (2007, p53). Hence, without a proper maturity IT organization, alignment alone can lead to fail (Marchand and Peppard (2008).These challenges form the basis of the refined research questions outlined in Section 2.7.

2.6 A Summary of the Gaps in Strategic Alignment Models

In this study I have chosen to focus on examining the aspects of Strategic Alignment and their importance on government IT projects. From literature discussed previously, the recognition of the importance of alignment has grown significantly as organisations strive to improve their performance through linkage between business and IT objectives. Different SA models have different focused attributes and thus there is no one model is able to solve all IT management issues (see Table 2-2). The implication is that these differences in models and attributes have consistently impacted on top management teams in public organisations where effective leadership is required to secure clarity in their IS/IT implementation in the light of their business objectives. However, without proper alignment of IT, it is unlikely that organisations will successfully deliver IT value to its stakeholders.

The issues of misalignment identified in literature are many and can be summarised as follows.

- *Public Sector Organisations (PSOs) are complex but still require SA*: The complexity of high-level integration in public organisations has not been adequately addressed by existing SA models (Weerakkody, Janssen et al. 2007). These models do not accurately reflect the enterprise-wide integration and collaboration that is necessary in government models. Public sector organisations are often more complex, have more stakeholders and are less concerned about financial issues than they are with public values which reflect their social, economic and political priorities (Dowse 2003; Gartner 2004; Weerakkody, Janssen et al. 2007). The scope of this research is to investigate SA strengths which improve productivity and facilitate service delivery within public organisations (Weill and Ross 2004; Guldentops 2011).

- *Previous research based on private organisations, not PSOs*: As Table 2-2 indicates, many of the strategic alignment models have been empirically investigated in the business sector (Elpez and Fink 2006).Alignment, however, is a concern not only for private organisations, but also for public organisations who have heavily invested in IT projects for public services purposes (Motjolopane and Brown 2004). The literature survey confirms previous findings that little research has been done concerning how to achieve IS success within the public sector (Brown, 1999 as cited by Garson, (1999).

- *PSO trends*: Since the existing power of IT in these private organisations encourages the consideration of dynamic consumer expectations, managing organisational transformations therefore becomes a strategic and integral part of overall organisational management activities. This trend is not exclusive to commercial institutions; the public sector is also adapting to similar managerial challenges (Tan and Pan 2003). In fact, public organisations are undergoing an identical transformation with the redesign of non-innovative bureaucratic governmental structures (Moon and Bretschneider 1997) to cope with a constant demand of quality public services (Guldentops 2004; Luftman 2011).

- *Another gap*: There is a need to define set criteria that identify which alignment perspectives contribute to organisational performance (Bhansali 2007).

- *SA value in PSO needs more input from stakeholders*: Strategic alignment does not efficiently account for other forms of government value (ITGI 2 007) which require more stakeholder engagement (Elpez and Fink 2006), such as sharing knowledge (Tremblay 2006) and a citizen interests focus (Cresswell, Burke et al. 2006). However, in order to allow IT to facilitate better coordination and productivity, it should be coupled with organisational changes and practices (Bhansali, 2007).

- *The continuing cost of IT failure is serious problem*: IT is an essential asset. The excessive and repeated cost of IT project failure in a government context calls for examining the conditions that pose a serious threat to the completion of the IT project (Sherer and Alter 2004; Stevens 2011) through hindering business/IT alignment (Luftman 2011).

In Table 2-2,the major gaps of these models and their limitations when applied to public organisations are summarised. The earlier alignment models (1960-1990) focused on factors affected by IT utilisation whereas later models looked to other factors in relation to IT as competitive advantages of organisation. Recently developed models emphasise knowledge and information as crucial to an organisation's success. Table 2-2 identifies the attributes of these models and assesses their weight as high or low attributes (represented in italics[11]) or inapplicability (underlined attributes) in the context of a public sector organisational context. The weighting of each factor varies over time and from one organisation to another (Ward and Peppard, 2000). Only high-weight value attributes are selected and used to develop an emerging conceptual framework. They serve to help in diagnosing and describing a comprehensive review of the concept of misalignment in greater details between business and IT strategy in a synthesised and conceptual manner that leads to developing the dimensions of a conceptual framework in Section 2.8.

---

11   Italic/underlined attributes are either considered more suited to private sector organisations or they are inapplicable in the book context and therefore were excluded to develop a conceptual framework. 'Governance' attribute, for example, was excluded as the concept of alignment is considered as part of a big picture of 'governance' Gartner (2004). www.gartner.com
          and that alignment has no value without good governance to be in place and addressed by top management. It's not that those perspectives and attributes which are excluded are not important, rather they are less relevant in the PSO context in the period of the book project.

TABLE 2-2: A COMPARISON OF SA MODELS FROM THE LITERATURE REVIEW

Models	SA Model Limitations (Section 2.5.1)	Engaging Stakeholders (Section 2.5.2)	IT Governance (Section 2.5.3)	Public Value Measures (Section 2.5.4)	High/Low Attributes	Applicability of SA models to PSO
Perspectives Affected by IT Utilisation: 1960–1990						
CSF-Rockart(1979)	Difficult to use – requires skilled analysts	Not addressed	Ineffective with lower-level management	Specific requirements not addressed	-IT resources (p) -MIS planning and activities (a) -Communication the role of IT to senior management (r) -Identify critical issues associated with implementation plan (t)	Partially, if properly used
Strategic Grid Method - McFarlan(1984)	No operational measures	With buyers and suppliers for the company's advantage	The earlier centralised concept of IT application	Profit-driven traditional ROI measures will fail in government (Gartner, 2004).	-The current and future IT application on business (c) *-Establishing barriers (d)* *-Competition based on cost (d)* *-Bargaining power with buyers (d)*	Reliance on financial measures that are not as well-suited for newer generations of IT application (Martinsons, Davison et al. 1998).
Other Perspectives in Relation to IT as Competitive Advantages of Organisation: 1991–2000						
The Balanced Scorecard - Kaplan & Norton (1992)	Focus on external factors only Financial term not suitable for PSO	Yes, but still inadequate for addressing stakeholders' contributions (employees, suppliers and community)	Yes, supported by establishing vision	The role of community (diverse stakeholders' contributions) receives less attention	-Developing performance measures (b) -Translating vision (j) (c) -Communicating strategic directions and goals (i) -Customers perspectives (e) -Internal perspectives (business process) (f) -Financial perspectives (value to shareholders) (g) -Allocating resources and establishing milestones (p) -Quality services (c) -Risk and control of IT (b) -Providing feedback and learning (i)	Is only capable of covering a portion of the needs of the public sector.
Strategic Alignment Model (SAM) – Henderson Venkatraman(1992)	Does not provide guidance to reach specific s trategic alignment goals		Yes, supported	Sharing internal and external information systems (a necessary attribute in this model) is necessary for alignment to occur (Chan and Reich 2007) due to the lack of coordination of different information systems in PSO(Weerakkody, Janssen et al. 2007)	-IT/business strategy (l) -Flexibility of IT-business integration (k), (p) -IT/business infrastructure (m) -Linkage between these interdependencies (j) *-Short-term decision* and long-term strategic planning (o) -Value management (o) -Governance (q) -Technological and organisational capabilities (o)	Yes, to some extent

Framework						
Management by Maxim - Broadbent & Weill (1997)	Technical criteria decision-based	Yes, for competitive positioning purposes	Addresses boards of directors' decisions on IT infrastructure	Too many options for configuring government IT investments makes decisions of top management based on technical criteria rather than in the context of long-term business needs which reflect their political priority, community needs, social and environment factors.	-Business/IT maxim (a business-driven IT infrastructure capabilities) (p) -Business maxim (deploy IT resources) (p) -IT maxim (decision on IT infrastructure services) (p) -IT infrastructure includes competitive positioning, improving cycle time, cross-functional processes, *utilising cross-selling*, new opportunities and channels to customers (m)	May be used
The Generic Framework - Maes(1999)	Too general	Embedded in the organisational environment	Supported, at strategy level where decisions are made on scope, core capabilities and governance	Embedded in the organisational environment	-Issues are considered at different levels -At strategic level (decision on scope, core capabilities and governance) (q) -At structural level (architecture and capabilities) (s) -At operational level (processes and skills) (w), (u)	Good support
The Integrated Architecture Framework - Cap Gemini Institute (2000)	Time-consuming and not straightforward	Yes	A coherent whole EA principles in PSO requires significant changes in organisational infrastructure	High integration in PSO is an excessive cost and is problematic even in developed countries. It requires harmonising coordination of different ICT infrastructures to reduce a legacy of rigid bureaucracy and time-consuming routine task	-Optimise all parts of business processes, information systems and organisation structures (a coherent whole of principles, methods and models) (u)	Requires training
The Social Dimension of Alignment Model - Reich &Benbasat (2000)	IT/business understanding requires a long time and a lot of effort to have a high level of shared domain knowledge culture	Yes, but still creates disadvantages in some stakeholder groups (due to the lack of a core method that effectively involves diverse stakeholder groups in policy process in PSO)	Business and IT executives still emphasis competition and contestability practices of private sectors. This disadvantages some stakeholders groups in government.	Yes, a core challenge for governments is the development of methods that effectively involve diverse stakeholder groups in policy processes, while avoiding unreasonable delays and unbalanced influence of unrepresentative stakeholder groups and invested interests (Bridgman and Davis 2004).	-Shared domain knowledge between business and IT executives (r) -IT implementation success (t) -Business/ IT executives communication (r) -Connections between business/IT planning processes (u) -*Short*- and long-term alignment (o)	Yes
SAMM – Luftman 2000	Simple to use	Yes	Yes, if addressed to top management	Metrics either not used or not evaluated by independent party	-Communication, value measurement, technology scope, partnership, governance, and skills (o) (s) -Descriptive and prescriptive aspects of alignment (P)	Can be easily misunderstood - different people see different thing in the same description

Public Value Framework Gartner (2004)	Requires professional	Yes	Coherence if well-addressed by top management	Hard to measure some of the intangible public returns	Safer community, citizen–focused, quality services, efficiency and effectiveness, user satisfaction (g), (b)	Yes, if well coordinated
A Knowledge-Based Theory of Alignment - Gartlan& Shanks (2007)	Diversity of stakeholders needs to be trained to become skilled human resources	Yes	Yes	The model shows an impact on short-term alignment (Hartung, Reich et al. 2000)	-Planning behaviour (business/ IT managers' participation in their plan) (w) -Top management knowledge of IT (r) -Contextual factor (knowledge focus and centralised decision of IT) (r) (p) -Quality of business/ IT planning and implementation problems in IT projects (u)	Yes, if well coordinated at different levels and supported by a top management level.

2.7 THE REFINED RESEARCH QUESTION

The gaps identified in Section 2.5 show that there is an insufficient understanding of how strategic alignment perspectives are deployed in IT governance contexts to ensure the success of IT projects in public organisations. Based on the factors identified in this chapter and in the context of our main research question: *'How is Strategic Alignment deployed in an IT governance context to ensure government IT project success?'*, this researcher aims provide insights on the perspectives and attributes applicable to public sector organisation through addressing the following sub questions.

RQ(a): What is the relationship between the SA maturity level and the success rate of IT projects?

RQ(b): Which Strategic Alignment attributes promote the success of IT projects?

RQ(c): Which Strategic Alignment perspectives promote the success of IT projects?

2.8 DIMENSION OF CONCEPTUAL FRAMEWORK: THE EMPHASIS OF CURRENT TRENDS

The concept of misalignment between business strategy and IT strategy as discussed in literature has been proven to be affected by many factors: cultural gap[12] (Grindly, 1992), reconciling the business/IT relationship (Ward and Peppard, 1999), measuring business/IT alignment (Luftman, 2003), shared domain knowledge (Hartung, Reich et al. 2000) and leadership accountability (ITGI 2003; Guldentops 2004). It is argued that these factors play a role in creating the misalignment between the business strategy and IT strategy (Shamekh, 2008).These factors are called 'attributes' in this book as previously shown in Table 2-2. They are overlapped, interrelated and sometimes repeated from one model to another.

In order to provide the ideal concept in developing frameworks for business/IT strategy in the public sectors, this section synthesises the alignment perspectives and attributes found in literature that are considered

---

12   A culture gap is an attribute that explains the problems that can exist between the IS/business function. Ward and Peppard (1999) considered it as a convenient description of the symptoms but not an explicit cause of the gap.

the most important. Accordingly, these perspectives play an essential role in creating the foundation of the SA model for the public sectors.

2.8.1 STRATEGY PERSPECTIVE

IT strategy plays a vital role in every organisation. Organisations cannot compete with each other without their unique strategies, so business/IT alignment is particularly important to maximise the benefits to the organisation that are gained from IT. The use of strategy by governments or public sector organisations is important not only to organise their processes but also to prepare emergency measures for whenever it they are needed.

Organisational strategy is also referred to as 'enterprise strategy' (Weill & Ross, 2004).For IT governance purposes, enterprise strategies should be clearly defined so that all the players in the process can easily use them. The Strategic perspective has three attributes that are described below.

Clarity of Direction: This attribute states that organisational objectives should be well communicated and understood via translating business vision to staff, communicating strategic directions and goals and long-term strategic planning attributes. Where there is no clarity, for example concerning as the objectives and goals of the business and IT systems, the direction of the SA being implemented in the organisation cannot be achieved. The clarity of direction as a criterion is compromised of the communication strategy, the mission statement of the organisation, the direction and the objectives (Luftman, 2000; De Haes & Grembergen, 2008). If there is clarity in the business/IT strategy of government projects, the functions and processes of the people are also well established, ensuring the smooth delivery of quality services.

Performance Measures: Performance measures include scope deliverables and the achievement of milestones, MIS planning, risk assessment and identifying CSFs associated with implementation of planned projects. These measures pertain to the activities within the organisation. Staff and their performance are an important aspect of the implementation of strategy. Here the importance of clarity in the direction and the alignment of the strategies in the objectives is evident. Lacking clarity in government projects also stresses the demonstration of the planned alliance between people, processes and the people with the processes that will lead to the expected outcomes (De Haes & Grembergen, 2008).

Quality of IT/Business Plan: A good IT/business plan has clear business and technical goals and requirements. It should contain sufficient and accurate details about the project's scope, costs and benefits, service legal agreements (SLAs), the allocated resources and established milestones and vendor-related issues if necessary. The quality of the project plan pertains to the internal strengths and proficiency of the organisation, as such plans help implement organisational strategy (Morgan & Schiemann, 1999). With the help of the quality measures, both small and large IT projects can be successful because the project plan and its strategy is focused on achieving its goals.

The three attributes - Clarity of Direction, Performance Measures and Quality of IT/Business Plan - are all dealing with measures of the quality of the IT/business outcome at an organisational level.

2.8.2 KNOWLEDGE PERSPECTIVE

In this book, knowledge means enabling people to quickly and accurately access the information they need so they can make the best business/IT decisions and determine the optimum course of action for better public value from IT investment. The knowledge perspective has shifted to different dimensions and garnered considerable attention from the community of IS/IT practitioners and academics. This direction has increased in public organisations to include a successful communication program (Carr, Folliard et al. 1999), empowering staff (Martinette and Dunford 2004), government information-sharing (NASCIO 2005), knowledge as power(Tremblay 2006),cross-boundary collaboration (Elpez and Fink 2006; NASCIO 2007), a Knowledge-Based Theory of Alignment (Kearns and Sabherwal 2007) and knowledge as strategy (Wang and Belardo 2009).

By implementing robust search capabilities and preparing to readily navigate abundant and expanding information resources, organisations ensure that tomorrow's employees benefits from today's expertise.

(Tremblay 2006, p.1).

Based on knowledge theory, a recent model linking strategic alignment between business and information technology developed by Kearns and Sabherwal (2007), revealed that senior managers' knowledge of IT affects strategic alignment by supporting the sharing of domain knowledge. In other words, the greater the knowledge of top managers, the more they support domain knowledge.

Results from a survey of 274 CIOs revealed that organisational emphasis on knowledge management and the centralisation of IT decisions affected top managers' knowledge of IT (Templeton, Lewis et al. 2002), affecting the planning behaviour of both business and IT managers (Johnston and Carrico 1988). This in turn affects business/IT strategic alignment. Achieving business-IT strategic alignment will increase quality of IT project planning and reduce problems during IT project implementation. Both of these will in affect the business impact of IT.

Knowledge in IT governance and alignment with business is highly dependent on communication. The communication framework has also been an area of debate since the development of IT into the management arena. According to a 2007 survey conducted by the Society of Information Management (SIM), only 45 per cent of chief information officers report directly to their CEOs. In 2008, that number decreased to 31 per cent. Luftman notices that with the organisations where the CIO does report to the CEO, the alignment maturity is clearly ahead of those where they do not (Luftman and Kempaiah 2008).

Knowledge is always the link in the strategy model. With the combination of strategy and knowledge, SA between business and IT is highly possible. Among all the aspects of SA, the aspect of knowledge is the core of the strategies because this perspective considers the ability of people to adopt the specific and appropriate knowledge in understanding the context of the business/IT strategic alignment. In connection to the government sector, the knowledge refers to the 'brain' of the organisation, for without the central program (the brain), the subordinates would not reach their full potential.The effect of knowledge in planning and implementation should clearly be addressed and the theories that reacted to them should be examined (Tallon and Kraemer 2003).Knowledge is composed of three attributes:

IT and Business Managers' Communication: Effective communication between IT managers and business managers should occur in the form of both formal and informal communication to facilitate project implementation. Such communication concerning business planning encourages coordination and collaboration. Like communication and collaboration, the managers of business and IT strategies should cooperate with each other to most effectively make plans and address the issues that arise in their fields. Through the help of the knowledge perspective, effectiveness in making decisions can be centralised. Also, the boundaries between IT and business management should be loose within the organisation. This will produce the mutual agreement and disagreement, which is healthy for organisations.

Organisational Emphasis on Knowledge: The application of IT in public sector or government agencies is a critical aspect that will challenge the skills and knowledge of the people who work in the organisation. In addition to communication at the upper management level, participation and sharing of knowledge amongst all the players in a system is vital.

The better the knowledge base upon which public policies are built, the more likely they are to succeed. (Riege and Lindsay 2006, p.28)

The influence of IT is not just due to IT's ability to deliver benefits, but also the values associated with them. Without careful planning, the potential of IT and its use with business functions cannot be fully realised. Therefore, rather than paying too much attention to the type of IT to be used, the public sector should pay attention to providing its leaders with opportunities to implement the training and development that will enhance the technological capabilities of the staffs as well as the capability to provide better linkage between the business and the IT strategies (De Haes and Grembergen 2008). It is also important that the people understand their organisational functions in order to improve their effectiveness and efficiency.

When organisational practices recognise the importance of knowledge-sharing, individuals are empowered to explore other opportunities for business/IT alignment (Luftman 2000; Al-Hatmi and Hales 2010). Ideally, this leads to improved communication and collaboration that strengthens the linkages of the processes and functions of IT.

Training: Ongoing training and workshops for staff and users are required as part of the knowledge perspective. Just as appropriate IT training is essential for successful project implementation, it is necessary for effective strategic alignment. Although the public sector is not-for-profit, it still needs to invest in IT systems that can satisfy its needs and, in order to realise the benefits of the IT systems, the government should demonstrate a plan that can facilitate technicians and end-users in utilising the IT system. Staff training, therefore, is crucial to success of IT projects and also their alignment. Through the participation of individuals, an organisation can evaluate the levels of individual staff and highlight the need for training across the organisation (Al-Hatmi and Hales 2010).

The root cause of difficulties in the engagement process of IT and business are the gaps in staff performance. The development of staff performances and the importance of the training analysis should be included in any effort to formulate the practices that positively affect the relationship of the business and the IT system that, in turn, achieve the business goals. Basically, when given proper training, workers and users adapt faster to the dynamic and changing world of IT.

2.8.3 DECISION-MAKING PERSPECTIVE

In the context of strategic alignment, decision-making is not about what specific decisions are made, rather it is a governance issue about systematically determining who makes each type of decision, who has input in the decision and how these people are held accountable for their role(Weill and Ross 2004). For these reasons, investment and budgeting for the decisions that concern IT is important (Brancheau and Wetherbe 1987). Good IT governance draws on corporate governance principles to manage and use IT to achieve corporate performance goals. According to this definition, governance determines who makes the decisions, whereas management is the process of making and implementing the decisions (Kearns and Sabherwal 2007).

IT Investments and Budget: Governments routinely spend huge amounts of public money in IT investments. Hence, making correct decisions about budget and IT expenditure is of immense importance in public sector organisations. The IT Investments and Budget attribute is a systematic approach to decision-making by whom and how based on the anticipated public value of IT, input from stakeholders and the degree of alignment with the organisation's ICT strategy.

For a systematic approach to decision-making, various structures should be used. A study by the Center for Technology in Government (CTG), which targeted building new capability for enterprise information technology investment decision-making in New York State, makes five recommendations (Pardo 2009):

- Establish an Executive Enterprise Governance Board (EEGB) to ensure alignment of enterprise IT decision-making with current state policies and strategic priorities;
- Establish an Information Technology Investment Board (ITIB);
- Adopt the CIO Council Charter as drafted by the CIO Council Action Team Co-Chairs;
- Establish a Technology Services Advisory Council (TSAC) to oversee the centralised IT services state agencies purchase from CIO/OFT; and
- Establish a temporary Enterprise IT Governance Implementation Committee with a responsibility for implementing the new IT governance structures and design a process for periodic review and assessment of how the new structure enhances the transparency, efficiency and coordination of the

state's enterprise IT investment decisions.

Prioritisation: A structured process of project selection is based on established criteria, such as risk status and cost/benefits analysis (Merkhofer, 2011). Ensuring that the appropriate business and IT participants formally discuss and review the priorities and allocation of IT resources is among the most important enablers/inhibitors of alignment. A well-designed priority system enables the organisation to identify the project choices that create maximum value for the resources available. This decision-making authority needs to be clearly defined (Sledgianowski and Luftman 2005). IT can be characterised as reactive to CEO direction, as the decision made by the CEO will affect the general strategy. Developing an integrated enterprise-wide strategic business plan for IT would facilitate better partnering within the organisation and lay the groundwork for external partnerships with customers and suppliers.

Prioritisation stresses a systematic approach to decision-making and is also somewhat focused on who is making the decision. This attribute elaborates on the idea of focusing on who will be the beneficiary of the decisions (Lange, Lee et al. 2011). In a government setting, structured decision-making is very important because the usual vertical hierarchies and integration of the government agencies can facilitate synergies across departments.

Information Systems are concerned with the process of collecting, processing, storing and transmitting relevant information to support the management operations in all organisations. In public organisations, budgets are constantly under stress and require transparency and justification, so an effective system that enables project prioritisation on the basis of strategic need to ensure that projects are selected by importance and their likely contribution to the organisation is essential. However, the success of decision-making is highly dependent on available information and on the functions that are the components of the process. Without information obtained through research, there are no alternatives to compare, and without a comparison of alternatives the choice of a particular course of action (the decision) is unlikely to yield the desired result.

Stakeholder Engagement: Effective decision-making processes must involve all stakeholders (Renaud and Waish, 2010). The degree of stakeholder engagement is based on the level of accountability and role of each individual. The principle of strategic corporate governance can be used to manage IT and achieve corporate performance goals(Ward and Peppard 2002).The decision-makers and decision-making processes will be discussed in greater detail in Section 4.6.1 in the context of IT governance structure. Decision-making occurs throughout each organisation's management hierarchy and varies from a tightly controlled centralised system to a loosely controlled decentralised system. All these systems are best used in the environments they were specifically designed for. In the case of public organisations, information systems work best where many stakeholders are engaged.

Decisions made on IT budget and investments across stakeholders is an emerging focus in the public sector (Elpez and Fink 2006; Valorinta 2011).The conclusion drawn from the study of three public organisation sectors conducted by Elpez and Fink(2006, p.220) was that the "success of an IS project is what stakeholders perceive it to be". Thus, stakeholder accountability becomes an important factor in the government setting, since it minimises the risks that may occur and thus encourages appropriate decision-making. Accountability and formal processes also require continuous monitoring by management. Overall, the extent and effectiveness of stakeholder engagement is determined by senior management, who have the authority and ultimate responsibility for managing change and its associated risks and benefits (Al-Hatmi and Hales 2010).

2.8.4 Enterprise Architecture (EA) Perspective

Enterprise architecture (EA) is defined as a conceptual framework for describing the architecture of a business and its information technology and the alignment between them. (Zarvic and Wieringa 2004). Architectural Design plays a significant role in business/IT alignment. Enterprise Architecture (EA) frameworks are used to control complexity and the high integration of information systems(Weerakkody, Janssen et al. 2007). EA, however, requires a continuous development in its IT adaption due to the rapid changes in today's public sector environment (Braun and Winter 2007). EA is a strategic tool or enabler to help align the government vision, business requirements, information systems and the public value of organisation. Hence, its role to public organisations is important(NASCIO 2003; Hjort-Madsen and Gotze 2004).

Enterprise architectures are generally seen as blueprints that identify the essential parts of an organisation (such as people, business processes, technology, information, financial elements, and other resources) and its information systems. (Ylimaki and Halttunen 2005/06, p.189)

The Integrated Architecture Framework (IAF) (Figure 2.2) was designed by Maes, Rijsenbrij et al. (2000) to support the role of IT in existing business processes. The role of IAF is to align IT with the business vision, eventually leading to IT-enabled enterprises. The key point here is that IT should be aligned with business goals and not planned according to the latter. When the IT management team links up as equal partners with the business executives, it dramatically increases the possibility of a positive outcome.

The components of business/IT alignment found in most Enterprise Architecture frameworks include business, information, application and technology architectures. Alignment here is defined as the development of information, application and infrastructure architectures in a way that completely supports the business necessities (Pereira and Sousa 2005). Figure 2.4 shows the role of IAF as illustrated by Maes, Rijsenbrij et al. (2000). The applications and technology include all the software and hardware used by the organisation to attain information that is deemed to be valuable. All of the programs in networks that are going to be used in order to get an end result are placed in form of a structure and the design of enterprise architecture.

FIGURE 2-4: THE ROLE OF THE INTEGRATED ARCHITECTURE FRAMEWORK(MAES, RIJSENBRIJ ET AL. 2000)

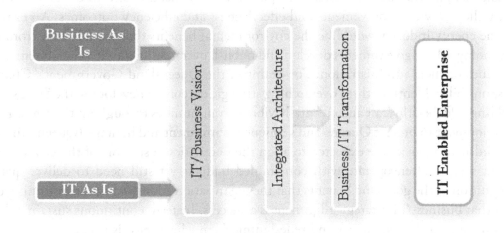

After designing a good architecture for the applications and technology of the enterprise, the necessary technical, manpower and training requirements can be obtained and put in place. The EA design plays a vital role in the business/IT alignment because it defines the interaction of the business processes with the overall information system. EA formalises the design of the information system and its support and interaction with organisational strategy. The following three attributes of Enterprise Architecture have been identified.

Aligned Technical/Business Solutions – It is important that the SA focuses on both the business objectives and the available solutions that the IT in the organisation can deliver. When IT solutions support the public organisation's business solutions, the delivery of government services can be more effective (Tallon and Kraemer 2003).

Application and Technology: The application of IT systems within the governmental setting is an important matter; however, the technologies that they use should be based on the requirements or the preferences of their functions that promote the effectiveness of the entire system. Therefore, management should research and determine the specific requirements for their organisation's need (De Haes and Grembergen 2008).

Risk Assessment: All IT carries a certain degree of risk, so government agencies should have well-developed risk assessment for the procurement and risk management plans for operations (Al-Hatmi and Hales 2010). Assessing the risks taken after deciding on the technology architecture is also part of the set-up in an organisation. For example, after a local council decides to store all its data digitally, questions should be asked about the security the new system. The security of the information is also put into jeopardy once the current technology starts to be used. Backing up and enforcing good security measures can be expensive and requires employing parallel IT professionals to design the structure in such a way so that discretion about certain information can be maintained. The money for these functions should be included as part of the budget. The goal is to make sure that no information is lost through any means. Therefore a secure system should be designed as an EA.

2.8.5 PUBLIC VALUE PERSPECTIVE

From a management perspective, strategy is defined as the statement of the specific intended results of enterprise's long-term direction, which is backed by values. Enterprise values are behaviours espoused by the enterprise, for example one stated aim of Deloitte is:" to see the Deloitte brand recognized as one of the most prestigious in the world in the terms of service quality, client preference, and being a place where people would like to work" (Deloitte 2005, p.35). In contrast, Value Based Management (VBM) is a management approach that ensures that an enterprise is consistently organised and operated to maximise shareholder value (Ittner and Larcker 2001). IT governance places the focus on shareholder values.

Moore (1995) introduced the concept of public value as the public sector equivalent of shareholder value. Value realisation includes the social benefits of corporate activity that are hard to define in monetary terms, such as public policy, service improvements and effectiveness and efficiency outcomes. An example of value realisation in the energy industry would be the environmental benefits that would come from developing a low-carbon economy. Many governments develop and publish overarching public value statements that define the strategic values of their administration. For example, the Queensland Government's Charter of Social and Fiscal Responsibility identifies three overarching strategic outcomes: developing the State's already strong economy; building safe, healthy, fair and culturally vibrant communities throughout the State; and promoting sustainable development to protect Queensland's unique environment and heritage (Queensland Government 2004). All subordinate agencies are expected to align their activities in support of these values.

Government sectors often operate with constrained budgets yet still need to deliver quality services. However, recognition of the goals and benefits that the organisation addressed in the beginning of the project does not mean that business/IT strategy alignment is a success. Instead, continuous sustainability is needed, which requires reviewing their strategies and redesigning them whenever it is needed.

There are three attributes identified that determine the value of IT in public organisations(Al-Hatmi and Hales 2010). The absence of these attributes (discussed below) might mean that the organisation is classified as having an improvised or unplanned maturity level of SA. In practical term, the importance of alignment in service delivery in public organisations is warranted and is considered a significant contributor to the overall quality of public customer service (Chang, Hsiao, et al, 2009).

Benefits to Organisation: Public organisations must plan for benefits that include the efficiency and effective delivery of existing and new systems. Organisations that propose new IT systems and applications should demonstrate how the new systems can make the organisation operation quicker, safer, easier and more effective.

Benefits to Public: Demonstration of public value promotes public interest and services delivery and therefore gains the trust of the public. The achievement of public value is the common goal of IT departments and public organisations and is the result of an effective SA strategy. Yang et al. (2010) discuss the values for public stakeholders who use government services, and argue that if citizens are satisfied, then public value has been achieved.

At an operational level, public organisations have to have the capacity to serve the public with the competitiveness and efficiency. Therefore, the most important operational goal is to target the public value that they set in their business/IT strategy (Khosrow-Pour 2000; Khaiata and Zualkernan 2009). The operation level within the public sectors might not be about market positioning but it is anticipated that the operation of the sector should be competitive enough not only to deliver quality in service but also to provide services for a large amount of people.

Economic/Financial Metrics: Traditional economic/financial metrics are needed to contribute to the overall estimation of cost-benefits analysis. Such metrics include Economic Value Added (EVA), Net Present Value (NPV), Internal Rate of Return (IRR) and Return of Investment (ROI) (ITGI 2008).

To conclude, the dominant strategic alignment perspectives in the literature are identified as follows:

- Strategy
- Knowledge
- Decision-Making
- Enterprise Architecture
- Public Value

Table 2-3 provides a summary of emerging attributes/perspectives and their descriptive framework for categorising key factors in ICT implementation in government with references to the literature. Next, a conceptual framework is developed that closes the gaps previously identified in the public context.

TABLE 2-3: DIMENSIONS OF A CONCEPTUAL FRAMEWORK

Attributes Identified From SA Models	Emerging Attributes Associated with PSO	Emerging Perspectives Associated with PSO
The current and future IT application for business (McFarlan 1984), translating vision and communicating strategic directions and goals (Kaplan and Norton 1992), long-term strategic plan (Henderson and Venkatraman 1992)	*Clarity of Direction*: it mainly focuses on organisational ICT-business objectives through translating vision and communicating strategic direction as also demonstrated in Luftman, (2000) and De Haes & Grembergen (2008).[13]	
MIS planning and activities (Rockart 1979), developing performance measures, risk and control of IT, and providing feedback and learning (Kaplan and Norton 1992)	*Performance Measures*: it is another important component of strategy perspective. Therefore, the performance measures include emphasises on project specific milestones, deliverables, and the risk assessment[1].	The three attributes, including Clarity of Direction, Performance Measures and Quality of IT-Business Plan, form the basis of organisational and IT strategy. They play a clear role to organise and direct IT-business processes and an emergency measures to be in place when needed (Norman Vargas, Leonel Plazaola et al. 2008). Collectively these attributes can be clustered altogether and called the ***Strategy Perspective.***
IT-business strategy (Henderson and Venkatraman 1992) internal perspectives (optimise all parts of business process) (Kaplan and Norton 1992; Maes, Rijsenbrij et al. 2000), Planning behaviour and quality of IT-business planning and implementation problems in IT projects (Gartlan and Shanks 2007)	*Quality of IT-Business Plan*: the third component included in the strategy perspective is *quality of IT/Business plan*. As noticed in Morgan & Schiemann (1999), this focuses on optimisation of all parts of business process for example from a concept plan to delivery of value IT project.	
Sharing domain knowledge and communication between business and IT executives (Reich and Benbasat 2000), planning behaviour (Gartlan and Shanks 2007)	*IT/Business Managers Participation*: this component is associated with the aspect of communication. Therefore, this element precisely focuses on the various informal and formal modes of communication between IT and business executives also observed in(Carr, Folliard et al. 1999).	All these attributes focus on sharing domain knowledge that facilitates business process and alignment of IT and business. They are therefore categorised as the knowledge perspective (Kearns and Sabherwal 2007). Therefore, ***The Knowledge perspective*** has increasingly given attention in public organisation to include cross-boundary collaboration (Elpez and Fink 2006; NASCIO 2007), empowering staff (Martinette and Dunford 2004) and government information-sharing (NASCIO 2005).
Providing feedback and learning (Kaplan and Norton 1992), communication the role of IT to senior management (Rockart 1979) connections between business/IT planning processes (Hartung, Reich et al. 2000), communication (Luftman 2000)	*Organisation Emphasis on Knowledge*: knowledge-seeking is obviously important for any organisation to update and enhance the required capacity. Thus, it focuses on Knowledge across organisation is a key factor for technological capabilities enhancement as demonstrated in (Riege and Lindsay 2006).	
Skills (Luftman 2000)	*Training*: similar to the knowledge the specific and effective training is another important component which emphasises on the on-going training, workshops and proper skilled human resources. This is also discussed in the recent past in (Al-Hatmi and Hales 2010).	

---

13   For details refer to De Haes & Grembergen, (2008).

Allocating resources (Rockart 1979; Kaplan and Norton 1992; Broadbent and Weill 1997), Centralised decision of IT (Gartlan and Shanks 2007)

IT Investments and Budget: The component of IT investment and budget assist in devising the decision-making perspective. Therefore, this factor emphasises on making government decisions regarding the budget and IT expenditure. It was also discussed in the previous literature on several occasions for example (Weill and Ross 2004; Pardo 2009).

Added value project management and value measurement (Henderson and Venkatraman 1992; Luftman 2000)

Prioritisation: this is another vital factor associated with the decision-making perspective. This factor focuses on those decisions which are connected with the concrete criteria of project selection, risk status and cost/benefits analysis. On the same lines it was also discussed in very recently by (Lange, Lee et al. 2011).

Customer perspectives (Kaplan and Norton 1992) Partnership (Luftman 2000)

Stakeholders: it is defined in different studies in a different way depending on the objective of studies. For instance, Elpez and Fink (2006) focus on sharing public policy on ICT investments in connection with the broader definition of stakeholders. Similarly, 'stakeholders' has been included in this book as one of the important component of the Decision-Making Perspective.

These attributes are related to one of the most important systematic approach in decision-making in regards to budget and government IT expenditure, added value selection of IT projects and diversity of public stakeholders. This is called the Decision-Making Perspective. **The Decision-Making Perspective** is an IT governance issue focusing on systematic decision on determining who makes each type of decision (a right decision), who has input to a decision and how these people are held accountable for their role (Weill and Ross 2004).

IT/business infrastructure, Flexibility of IT/business integration (Henderson and Venkatraman 1992)

Align Technical/Business Solutions: This factor constitute to the enterprise architecture perspective. And it focuses on the dynamics of architecture and infrastructure of a business and IT. Similarly, it is explained in different studies for instance(NASCIO 2003; Hjort-Madsen and Gotze 2004).

The current of future IT application on business (McFarlan 1984), IT/business Maxim (Broadbent and Weill 1997), Architecture and capabilities (Maes, Rijsenbrij et al. 2000)

Application and Technology: This as another factor of enterprise architecture perspective focuses on applicability of IS/IT in governmental setting and its ICT strategy and objectives. Similarly it is also discussed in (Weerakkody, Janssen et al. 2007).

Such attributes deal with the whole integration parts of technology infrastructure of an organisation such as people, business processes, technology, data, information systems and resources (Ylimaki and Halttunen 2005/06, p.189). This strategic integration attributes tool is called **Enterprise Architecture Perspective** and is used to align government vision, business requirements, information systems and public value of organisation (Zarvic and Wieringa 2004). Its role in public organisation is considered important (NASCIO 2003; Hjort-Madsen and Gotze 2004).

Risk and control of IT (Kaplan and Norton 1992), IT implementation success (Reich and Benbasat 2000), value measurement (Luftman 2000)

Risk: This is one of the important factors which is affected by organisational enterprise architecture. Therefore, it emphasises on risk/value management plan and security measures that should be in place in any emergency as discussed in (Al-Hatmi and Hales 2010).

Quality services and internal perspectives (Kaplan and Norton 1992)	***Benefits to Organisation***: the angle of benefits to the organisation is important in connection with the public value perspective. This primarily focuses on the internal benefits such as efficiency and effectiveness as also observed in (Khosrow-Pour 2000; Khaiata and Zualkernan 2009).	The all these three attributes can be clustered around the single perspective that is known as ***Public Value Perspective.*** Broadly, these attributes relate to public
Communicating channel to customers (Broadbent and Weill 1997) Value measurement (Luftman 2000) Customer perspectives and Quality services (Kaplan and Norton 2004)	***Benefits to Public***: benefits to the public are also associated with the public value perspective. This factor precisely focuses on external benefits to citizens and end-users (Moore 1995; Yang, Shen et al. 2010)	value of IT investments which often reflects not only ROI in financial term, but rather other forms of benefits in public such as efficiency, effectiveness, quality services, and political returns (Khaiata and Zualkernan 2009; Yang, Shen et al.
The current and future IT application on business in financial term (McFarlan 1984)	***Financial Metrics***: The third important component of public value perspective is the financial metrics. This precisely corresponds to the benefits that can be broadly accounted economically (ITGI 2008).	2010).

2.9 CONCEPTUAL FRAMEWORK

The literature review discussed the aspects of strategic alignment and the gaps associated with the suitability of current models for addressing SA within a government context The discussion proposes that strategic alignment in a public organisation is unique because the public sector's goals of achieving value for money while at the same time achieving its public value objectives are different to the financially-focused goals of private enterprises. Through reviewing the related studies pertaining to the SA, the study can see a great possibility for initiating SA as part of operations in the public sector.

The five focus areas from the various models of SA have been identified, discussed and categorised under broad common categories for investigation in a public agency context. In this research these were considered to be the major contributors to project success, however the categories are not exhaustive, due to the diversity of the implementation environments. This results in a rich picture of the ICT implementation experience that helps to identify possible solutions for the ongoing problem of project failure. It is expected that SA perspectives will contribute positively to value realisation with government IT projects. With this aim in mind, and based on themes derived from literature review and research questions addressed, a conceptual framework is proposed. The model builds on the rich history of promoting SA in the business operations, but modifies this to ensure its applicability in the public sector. It is an important model as it has the potential to reliably enhance the performance of the public organisations. Figure 2-5 illustrates the conceptual model developed as part of this research.

FIGURE 2-5: CONCEPTUAL FRAMEWORK

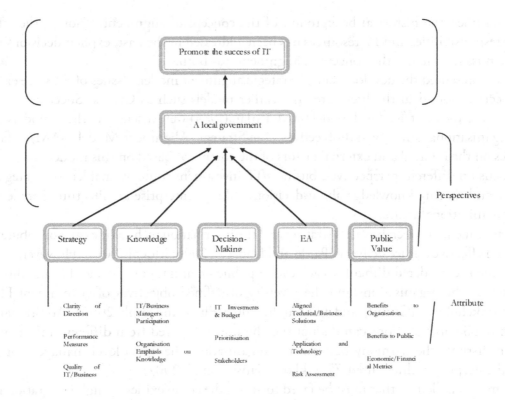

The model in Figure 2-5 provides a comprehensive illustration of the dimension of the five perspectives applied and considered as the contributing factors towards the strategic alignment perspectives. The SA perspectives that are expected to influence the success of IT projects in public organisations are Strategy, Knowledge, Decision-Making, Enterprise Architecture and Public Value. The framework explains the most efficient way in which SA in the public sector can be achieved. Based on the framework, there should be formal standard and mechanisms that can be used in order to help the organisation generate the success.

The above SA perspectives are in turn influenced by their characteristics (called 'attributes') as indicated in the model. As we have seen earlier (Section 2.3.9), while Luftman's alignment model consists of twelve components of alignment, the conceptual model developed in this research consists of fifteen attributes of Strategic Alignment. Luftman's measurement maturity idea of alignment was adopted in this research to describe the characteristic attributes of SA perspectives that are consistent with the literature. Decision-making, for example, is highly influenced by IT investments and budget, prioritisation of projects and stakeholder engagement in public policy and the process of the life cycle of a project's implementation. This is an effective concept since it contributes to the maturity level of the SA perspective and, through the help of their attributes, all of the processes of project implementation are taken into account. In addition, there is the potential that the two strategies (business and IT) can be synchronised, meaning that the attributes that are under each perspective are also examined. It is important that the framework is used in organisations, because it represents the linkage between the business and IT strategies in order to generate the expected public value.

Moreover, the success of IT projects is actually based in the high maturity level of the perspectives outlined above and since the business environment (including the public and government sectors) is engaged in a rapid change of pace in regards to IT, these attributes can be adjusted to make the criteria suitable for the current trends of government policies. In this strategy, policy-makers can optimise the value of their IT projects. If these domains are approached holistically and IT governance is implemented properly (De Haes and Grembergen 2008), many of the limitations in public organisations will be overcome.

2.10 Conclusion

This chapter outlined the historical background of the concept of alignment. Though the relationship of management responsibilities and IT resources has been addressed in the past, explicit decisions regarding IT activities were missing. Hence, the concept of alignment was born.

This chapter presented the development of strategic alignment models. Issues of SA were examined from various perspectives found in the literature. The earlier models such as Critical Success Factors (CSF) by Rockart (1979) and the Strategic Grid model by McFarlan (1984) were discussed. These models focus on how IT links to organisational strategy and objectives. The Strategic Alignment Model (SAM), a famous earlier model, focuses on the internal and external factors of the company. Based on this model, many other models emerged to focus on different perspectives: business/IT investment strategy, enablers of strategic alignment, accountability and sharing knowledge about decision-making, enterprise architecture, people engagement, process and organisational issues.

Effective alignment between business strategy and the information technology can contribute significantly to efficiency and effective service delivery (Pardo and Dadayan 2006; Mocnic 2010). However, service delivery through IT value is considered difficult to achieve in public organisations (Siau and Long 2004) due to the complex nature of the organisations and the diversity of defined objectives of government IT investment (interests) by stakeholders (Griffith and Gibson Jr. 2001; Elpez and Fink 2006). Additionally, a lack of integration in public organisations can disrupt the alignment required from different information systems, which in turn disrupts the harmony between the organisation, the upper-level management strategy and organisational enterprise architecture (Weerakkody, Janssen et al. 2007).

There are many challenges that must be faced to attain the complexities of high integration necessary for achieving business/IT success in public organisation, including the high cost of IT expenditure (Cresswell, Burke et al. 2006), alignment as a process (corresponding to the rapid change of technology), which makes it harder to consistently determine which alignment perspectives contribute to organisational performance and a lack of stakeholder engagement (NASCIO 2007). To ensure strategic alignment, a formative evaluation model should include training upper-level management, engaging stakeholders' cross-boundary collaboration and well-defined IT objectives through an IT/business plan that focuses on public return of IT investments.

Hence, a conceptual framework was developed to address those gaps. The challenges previously outlined can be surmounted through increased efficiency and effectiveness combined with innovative applications of IT, as well as awareness of both IT and organisational strategies, leading to more strategically oriented approaches for planning and management. While perspectives affecting the success of IT projects have many dimensions, this study will focus only on those dimensions related to high-level executive management in public sectors in an IT governance context. Such concerns are useful to measure costs, benefits, social intangible benefits, political returns and the efficiency of government services. All other dimensions related to operational levels and end-users will only be briefly discussed. The next chapter addresses the methodology used in this research.

3. RESEARCH METHODOLOGY

3.1 GENERAL REMARKS

This study adopted qualitative research methods to explore how SA perspectives are deployed to ensure the success of government IT projects. To fully address the nature and context of the research problem, a local government case study and in-depth qualitative research were used. The aim of this qualitative research is to "engage in research that probes for deeper understanding rather than examining surface features" (Johnson 1995, p.4). The goal of this approach is to understand individual IT project cases that bring to the forefront voices that have previously been marginalised (Perl and Noldon 2000).

This research analysed data obtained from interviews, government documents and participatory observation. These data sources proved insightful in understanding and explaining social phenomena in the real context of a local government. The methodology is supported by Myers' observation(1997) that there has been a general shift in IS research away from technological issues towards managerial and organisational issues.

This chapter is organised as follows. This section (3.1) provides general information about Chapter 3 followed by the chapter structure. The reasons for adopting the case study methodology are discussed in Section 3.2. Section 3.3 briefly describes the unit of analysis, while Section 3.4 provides the approach used in data analysis, Section 3.5 gives an overview of the research design, which includes sample selection procedures, data collection, observation, archival records and interviews. In Section 3.6, precautions undertaken with regard to ethical considerations, including ethical matters to protect the participants, are stated. The validity of research design, including construct validity, internal/external validity and reliability, is addressed in Section 3.7. Limitations of the methodology used are outlined in Section 3.8. The concluding section, 3.9, summarises the chapter.

Figure 3.1 diagrammatically represents the book research design. The circle with dark bold line represents the location of this chapter in the book.

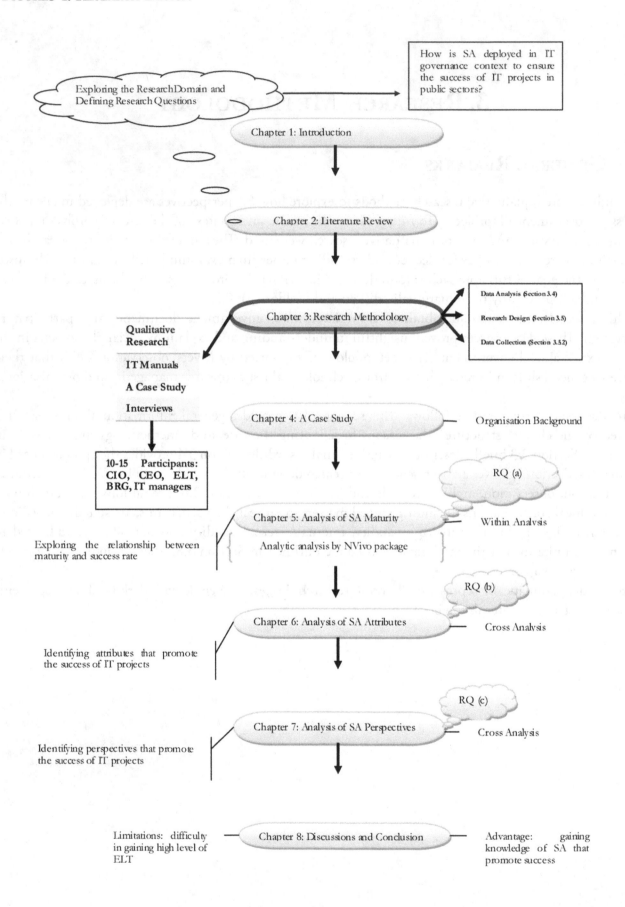

3.2 Reasons for Adopting the Case Study Methodology

This research used the case study method because that method is the best suited to exploring and understanding the deployment of SA aspects and their impact on IT projects in government settings. This method allowed the researcher to investigate the pattern of historical data of IT projects from public documents, which offered detailed information about various aspects of other IT projects, both current and past. Using a case study also allowed the researcher to observe organisational behaviour in the real-life context in which contemporary events occur. In this way, the researcher was able to ask 'how' and 'why' questions about the purposes of a social phenomenon in government settings without interfering or controlling its behaviour. Such a method is well supported by other studies (Yin 2003). Darke, Shanks and Broadbent (1998) state that the case study is the most widely used qualitative research method in IS research, and is well suited for understanding the interactions between IT-related innovations and organisational contexts.

Goode and Hatt(1952) define a case study as a way of organising social data so as to preserve the unitary character of the social object being studied. In other words, it is an approach that views any social unit as a whole. According to this definition, a social unit, a real individual, social event or group of people who are relevant to the scientific problem being investigated are treated as a whole whose characteristics are kept together. A case study, sometimes called a 'monograph', means studying only one event, process, person, organisation unit or object. Yin (1988) has defined a case study as an empirical inquiry that investigates a contemporary phenomenon within its real-life context, especially when the boundaries between phenomenon and context are not clearly evident.

Yin (2003) supports the use of a single case study because it provides an appropriate methodology for studying the phenomenon in the context in which it occurs. Case study research offers a unique, revelatory or critical case in support of or against a well-formulated theory. Government IT projects and the related IT governance context are well documented and can be studied by reviewing a variety of historical materials. Major government IT projects require effort and time to both implement and judge their failure/success, and the processes involved are often not publicly accessible. Therefore, since a longitudinal research approach is required, the single case study methodology is the most appropriate when a holistic, in-depth investigation is necessary. Thus, the case study in this research is defined as an integrated system bonded by time and place (Stake 1995). In the explanatory case study, I address the purpose of explanation, what is to be explained and the criteria by which the explanation will be judged successful. Hence, the selection of the case study was the most suitable, and this method was adopted for the following reasons:

- The subject under investigation was a public organisation with significant experience of government IT projects;
- Entry and maximum access were possible to maintain continuity of presence for as long as necessary (Marshall and Rossman 1989), as trust had already been established between the researcher and Council staff;
- There was a high probability that a rich mix of processes, structures, interpersonal interactions, IT projects, priority and decision-making that were to be a part of the research questions would be present;
- By avoiding poor sampling decisions, data quality and credibility of the study were reasonably assured (Khaiata & Zualkernan 2009);
- The phenomenon was examined in a natural setting (Eisenhardt and Graebner 2007);
- The case study approach was useful in the query of 'why' and 'how' (Gibbert and Ruigrok 2010);
- The focus was on contemporary events (Benbasat, Goldstein et al. 1987);
- The phenomenon was broad and complex;
- Data in this case study method was collected by multiple means including interviews, document analysis and observation;

- The existing body of knowledge was insufficient to permit the posing of causal questions;
- A holistic, in-depth investigation was required;
- A case study methodology added the theoretical value and explanatory efforts of quantitative work (Chan, Huff et al. 1997); and
- A phenomenon cannot be studied outside the context in which it occurred (Bonoma 1985; Benbasat, Goldstein et al. 1987; Feagin, Orum et al. 1991;Yin 2003).

Therefore, as our research focuses on SA perspectives and the success of IT projects in organisations and the focus of research has shifted from technical issues to organisational processes and interactions, the case study methodology is particularly well suited in this situation (Benbasat, Goldstein et al. 1987). The case study approach enabled the researchers to explore the queries in the IT governance context (structures, process and relational mechanisms), the frameworks used, IT project characteristics, project documentation (such as concept plans, charters, business cases, value realisation plans and the post-implementation review), SA aspects, risk assessments and maturity measurements, all of which provide valuable and critical insights. The approach was particularly relevant to this research as it allowed for understanding emergent relationships with various SA perspectives as well as providing a good understanding of the dynamics underlying these relationships to project outcomes in the public sector.

It is important to remember that case studies, however, are not simply a narrative account of an event or a series of related events; they must have a theoretical aim and involve units of analysis of a particular phenomenon that have been assembled with the explicit view of drawing theoretical conclusions from it in support or against such theory (Mitchell 1983).

Table 3-1 provides examples of some cases focusing on SA research, mostly in public organisations. This study will follow the same path of case study known as a 'theoretical sampling' also called the purposeful sampling approach.

TABLE 3-1: EXAMPLES OF SA RESEARCH USING THE CASE STUDY METHOD

Authors	Journal/Book	Case Description	Research Issue
(Carr, Folliard et al. 1999)	Bell Labs Technical Journal	Success of 5ESS@-2000 Switch project	Communication channels
(Grant 2003)	Journal of Information Technology	A case study of Metal co	Strategic alignment and enterprise implementation
(Tan and Pan 2003)	European Journal of Information Systems	A case study of IRAS, Singapore	Managing e-transformation in the public sector
(Motjolopane and Brown 2004)	SAICSIT	A case study of the South African Government's IT systems	Strategic business/IT alignment
(Byrd, Lewis et al. 2005)	Journal of Information and Management	An empirical examination of the impact of alignment on payoff of IT investments	The impact of SA on IT investment
(NASCIO 2005)	Bureau of Justice Assistance	A case study of government IT project management assessment in the U.S.	Government information-sharing
(Ylimaki and Halttunen 2005/06, p.189)	Information Knowledge Systems Management	Applying Zachman framework in the small EA projects perspective	Alignment between business vision and IS
(Aurum, Wohlin et al. 2006)	International Journal of Software Engineering	An industry-based case study	Decision perspective on IT software projects
(Cresswell and Burke 2006)	Center for Technology in Government	A case study of Washington State Digital Archives	Delivering IT value to the public
(Burke and Cresswell 2006)	Center for Technology in Government	A case study of the Austrian Federal Budgeting and Bookkeeping System	Delivering IT value to the public
(Cresswell and Burke 2006)	Center for Technology in Government	A case study of the mobile data communication and web-based transactions of Israel's Merkava Project – government IT initiatives	Delivering IT value to the public
(Dawes, Burke et al. 2006)	Center for Technology in Government	A case study of Pennsylvania's Integrated Enterprise System	Delivering IT value to the public
(Pardo and Dadayan 2006)	Center for Technology in Government	A case study of Service New Brunswick, Canada	Delivering IT value to the public
(Elpez and Fink 2006)	Journal of Information Science and Information Technology	A case study of IS success in the Australian public sector	Stakeholders' perspectives and Emerging Alignment Model
(NASCIO 2006)	Government report	A case study of increase IT implementation success by PM in the U.S. government	Challenges and opportunities for IT PMOs
(NASCIO 2007)	Government report	A case study of CIOs collaboration in US government	Cross-boundary collaboration
(Fardal 2007)	Information Science and Information Technology	An exploratory case between ICT users and managers' alignment in Norway	SA between ICT managers and ICT users
(Gist and Langley 2007)	Journal of Research Administration	A case study of a multi-national clinical trial	Application of standard project management tools to research
(Oh and Pinsonneault 2007)	MIS Quarterly	110 firms	Strategic value of IT
(ITGI 2007)	ITGI publication	A police case study in Australian government	Outcomes of IT investment
(Tzeng, Chen et al. 2008)	International Journal of Production Economics	Five case studies from Taiwan health industry	Business value of RFID
(Firth, Mellor et al. 2008)	Australian Health Review 2008	A case study of the negative impact on nurses of lack of alignment in Australia	The negative impact of misalignment of IS with public hospital strategy goals
(Raven 2008)	Bulletin of Science, Technology and Society	Two case studies of 27 emerging energy projects in Europe	Alignment between local expectations and local solutions
(Weilbach and Byrne 2009)	A conference proceeding	A case study of an open source enterprise system in South African public sector	Public policy and IS innovations alignment
(D'Angelo and Abramo 2009)	Science and Public Policy	The case of four high-tech sectors in Italy	Alignment between knowledge public supply and industry demand
(Wang and Belardo 2009)	Journal of Information Science	A case study of two energy companies in Taiwan	The impact of knowledge on effective crisis management
(Mocnic 2010)	Managing Global Transitions	A company-wide perspective of achieving value for customers	IT/business communication
(Yaghoubi 2010)	Journal of US-China Public Administration	The rural development in developing countries	IT-based rural development infrastructure
(Lobur 2011)	Insights: Project Management	A case study of the success of a COTS caseload management systems in the U.S.: Commonwealth of Pennsylvania	Projects failure even despite strong project management

3.3 Unit of analysis

The unit(s) of analysis is an important component that must be considered in any research design. Without a clear design of the unit(s) of analysis, the researcher would not be able to limit the boundaries of the study (Pare 2004). Based on literature review, research question(s) and research design, this research developed a conceptual framework of SA perspectives: that is, a focus area, in which a single unit of analysis of a holistic case study was framed. These perspectives and their impact on government IT projects were examined, based on investigation suggested and deriving from the main research question. Hence, specifying the unit of analysis is important in order to understand how this case study relates to a broader body of knowledge.

Based on the literature review, research question(s) and research design, our research has three main components of analysis: IT governance, SA perspectives and government IT projects. IT governance consists of structures, processes and relational mechanisms, all of which occur at a high level of management practices in a local government, the chosen case study of this book. In SA perspectives, dimensional factors are investigated in the context of IT governance and their impact on government IT projects. These perspectives are strategy, knowledge, decision-making, enterprise architecture and public value. Based on these components, the main unit of analysis for this study is SA in a government IT project context as suggested by Yin (1988).

3.4 Data Analysis Approach

As stated earlier, this study followed a qualitative approach to analyse data and is consistent with the overall case study approach (Darke, Shanks et al. 1998). When analysing data, precautions must be taken in the way data is segmented into units and rearranged into categories (themes) for the purpose of facilitating pattern-matching, comparison and in-depth insight between segment data (Eisenhardt 1989; Strauss and Corbin 1998). Hence, the NVivo analysis tool was used. The NVivo package tool served as a rapid and effective way for retrieving, clustering and organising coded segments to a particular query or theme. The detail of this analytical approach is given below.

Gathering your source materials: Qualitative data is often rich, complex, and vivid. Sometimes it is difficult to reduce sources of data sufficiently to make sense of events or manage an increasing body of data. To minimise this challenge, the data analysis steps approach outlined in Figure 3-2 was adopted because it is a well-described accepted approach within IS research (Miles and Huberman 1984)and it provides a systematic approach to gather, reduce and manage data that facilitates analysis rigour of the study. As Lyn (2005) states, the challenge is not so much making which is but rather making useful, valuable data and relevant to the question being asked.

Figure 3-2: Data Analysis Steps

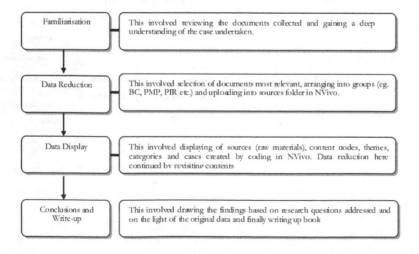

The relevant documents analysed, including interviews gathered from a local government, were imported into NVivo in a folder called 'Internals'. These documents were then clustered into groups (subfolders) according to the documentation process of lifecycle of projects such as BC, PMP, and PIR. Interviews transcripts of 11 participants were also stored here in a separate folder called 'Interview'. Appendix L shows that eight folders were created in this research and each folder contains several documents. Arranging documents in this way made it easier for the researcher to rapidly retrieve or relate queries to a particular document.

Creating Tree Nodes: The themes of this research were already known to the researcher due to their theoretical relation to the research question and design. They were then developed and explored, thus creating nodes into NVivo. A node is a collection of references about a specific theme, person, or other area of interest. In this research, for example, the 'Strategy' perspective is a parent node and consists of three attributes known as child nodes. Organising nodes into a hierarchical structure (called 'tree nodes') allows the researcher easy access to them, like a library catalogue. As the research progressed, the following nodes had been slightly refined to capture the meaning intended:

- 'Public value' instead of 'Value realisation'
- 'Aligned technical/business solution' instead of 'Align technical solutions with business solutions'

In addition, a node called 'Report of CIO', was deleted while another, 'Systematic approach of decision-making by who and how', was merged into another node, as they proved not to be sufficient enough to be stand-alone nodes (themes). Figure 3-3 provides data structure approach that was inductively used to analyse data via the naturalistic inquiry method (Lincoln and Guba, 1985). Such method helps to develop a rigorous strategy for analysing qualitative data to simultaneously assist and constant comparison technique in determining the sampling and content of data collection efforts. The method identifies initial concepts within the data and groups them into categories known as child nodes (second order themes). Due to relationships amongst these categories, further inducing second order themes into aggregate dimensions was carried out. Gathering similar themes into the aggregate dimensions makes up the basis of the emergent model and was part of a naturalistic analytic procedure (Locke, 1996) for obtaining a solid grasp of the emerging theoretical relationships. The full structure of tree nodes of this research is also illustrated in Appendix L.

FIGURE 3-3: DATA STRUCTURE

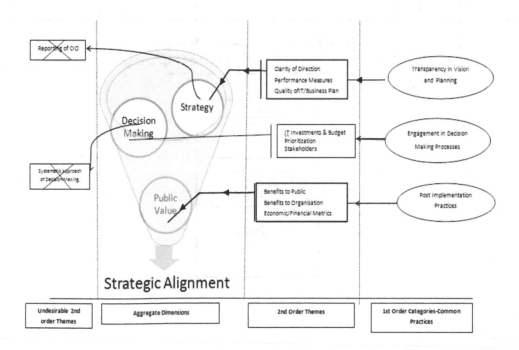

Coding: Gathering the references by reading through documents and interviews and categorising information into the relevant nodes is called coding. Figure 3-4 below illustrates the process of gathering material for a project from multiple sources: namely BC, PMP, interviews, and ICT Strategy, into a node called 'Clarity of Direction', thus allowing the researcher to explore the content of these attributes in one place. As the coding progressed, it was possible for the researcher also to run a query and retrieve the content of 'Clarity of Direction' node cross projects and build a query of the possible connection between this node with other nodes such as the 'Decision-Making' perspective or with only cases (projects) with low maturity level or with only the cases (projects) that were considered big size[14]. In this way, the coding was a way of abstracting different source data to build a greater understanding of the ideas and concepts being undertaken. The selection of four coders, two professors, my PhD colleague and myself, helps to develop more accurate and robust codes (Irava 2009). Inter-rater reliability measurement was carried out using Cohen's Kappa coefficient formula of inter rater to indicate the strength of agreement in coding among the raters (Fleiss, J. 1971; Sim, J. and Wright, C. 2005; Strijbos, J.; Martens, R.; et al 2006). Refer Appendix M for the calculation of the Cohen's Kappa coefficient.The process of coding outcomes was also carried out via peer review internal seminar and via conference presentations of the data and coding. Later it helped to find hidden patterns, building theories and seeking relationships between categories.

FIGURE 3-4: PROJECT CODING METHODOLOGY

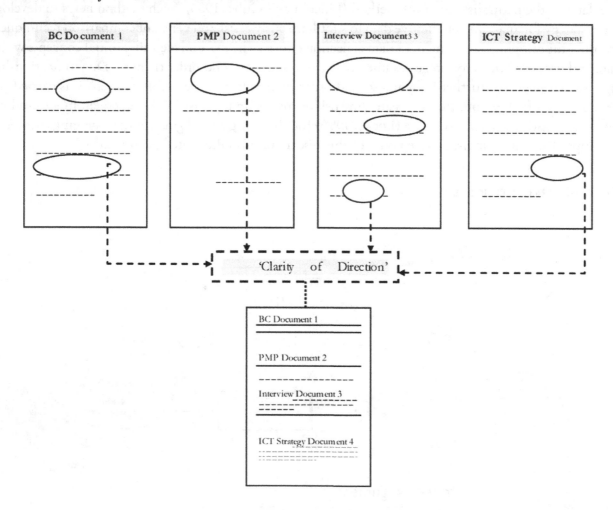

---

14   A project is considered 'big size' if it has a budget of more than AU$ 2 million.

As coding the content of data progresses may change at different stages of the analysis, skill with coding and an overall purpose are required. To meet these requirements, four approaches to coding were used in this research which contributed differently to the process of analysing the data (Strauss and Corbin 1998; Babbie 2001).These approaches are explained as follows.

Descriptive coding is the process of coding information that describes the cases in the research. These cases (14 projects in this research) were further classified into attributes and attribute values. 'A Summary of a Case Project Characteristics' (Table 4-1) was obtained as a result of the descriptive coding process. Appendix N provides an example of a case (referred to Project 4 in this research) and its attribute value.

Topic coding is the process of selecting contents and assigning them to the topics and categories they related to (Boyatzis 1998). It is important to see all information about these topics, categories, themes and nodes to enhance further analysis. The coding of contents from various sources to the 'Stakeholders' node can be an example of this.

Analytic coding is the process of interpreting and reflecting on the meaning of the data to arrive at new ideas and categories. It was the necessary way of gathering material that need to be rethought and refined to gain new insights and a growing understanding of the categories undertaken (Krippendorff 2003; Yin 2003). The NVivo8 allowed the researcher to explore the nodes contents using coding query or matrix coding query and save the results as new node (concepts). Miles and Huberman (1994, p.38)note that "issues of instrument validity and reliability ride largely on the skills of the researcher".

Query coding is the process of running a query by analysing, searching, filtering, and most importantly selectively interrogating the data, thus discovering and exploring patterns, testing hunches, developing and building theory. Running query occurred once nodes had been created to represent the themes, topics, the categories in the data and the contents of sources to those nodes had also been coded.

3.5 OVERVIEW OF RESEARCH DESIGN

3.5.1 SAMPLE SELECTION PROCEDURES

In every research design, the rationale for selecting the sample and its size is crucial. The sample for this study followed a purposeful sampling approach. The reliability of a sample depends on how well it represents the characteristics of the entire population. Population is defined as the aggregate of all cases that conform to some designated set of specifications (Nachmias and Nachmias 1996). A sample is considered representative of the population if the analyses made using the researcher's sampling units produce results similar to those that would be obtained had the researcher analysed the entire population. In this section, a sample methodology and the size of population of a local government case study are identified and justified. The characteristics of the sample are also provided.

In the local government case study, the sample is purposive; it comprises of 11 IT professionals at a high level of ICT management who were interviewed. The unit of the sample investigated comprises 14 IT projects in this public sector organisation.

The characteristics of the IT professional sample can be summarised as follows:
- They are the core source for ICT governance knowledge and practices;
- They are at a high level of decision-making concerning IT investment and related IT activities;
- They are the owners who are responsible for clarifying the direction of business and IT strategy and their priority. They are sometimes the end-users of the systems being implemented; and
- They adopt frameworks (based on COBIT and ITIL) that support their corporate strategy.

Patton (2002) states that the logic and power of purposeful sampling lies in selecting information-rich cases from which one can learn a great deal about issues that are centrally important to the purpose of the research's in-depth study. The study therefore involves a small sample of IT professionals who held the knowledge to provide relevant information needed.

Hence, a non-probability sampling design was used in this research. This design does not give every population element a chance of selection. Rather, as studying a whole population may be slow, tedious, expensive, unnecessary or impossible, the researcher purposively selected sampling units subjectively in an attempt to obtain a sample that appears to be representative of the population. A sample derived through such sampling methods is called a purposive or judgment sample. This means that from the IT professional sample participating in this research, only those who explicitly displayed interest in problem-solving issues and were likely to provide relevant rich data for extending emergent theory were chosen. Purposeful sampling is ideal for exploratory, qualitative and case-based research. The selected rich information helps in addressing the key research problem.

3.5.2 DATA COLLECTION

The use of multiple data collection methods (also known as triangulation) is considered useful when using case studies (Eisenhardt and Graebner 2007). Using various sources of empirical materials provided the rigour and breadth of understanding to fully address the research problem in a case study. The fundamental techniques used in this research for gathering information are observation, archival research and documentation as well as open-ended and in-depth semi-structured interviews. The primary data source was interviews. Secondary data sources included government documents and public material, websites, meetings minutes and correspondence with participants. Interview quotes, due their richness and description, are used in this book to highlight case features. This section outlines the methods adopted for this research.

3.5.2.1 OBSERVATION

Observation is the systematic description of events observed in a particular setting and is a vital technique in empirical research. The purpose of observation is to witness factual situations and perceive reality without intervention (Buckley, Buckley et al. 1976). A good researcher is a highly skilled observer who has the ability to observe different aspects of social interaction connectivity and casual relationships, identify relevant information from irrelevant and derive theories capable of further testing. There are two methods of observation: participant and non-participant. This research is based on participant observation, which is most appropriate with small populations, access to IT activities and frequent occurrence of events in a short period of data collection. Hence, the researcher provides a meaningful and more coherent picture of a single local government setting.

During field visits, data collection activities and interviews, the less formal method of observation was used. Observational evidence is useful in providing additional information and gaining clarity about the topic being studied. In addition, the researcher assumed a variety of roles, such an internal observer and participated in events such as presentations, taking minutes at meetings and the reviewing of value realisation plan documentation. Such participation enabled the researcher to gain access to IT project processes, government ICT strategy and decision-making priorities as well as connecting with business owners. In short, the observation method enables the researcher to perceive reality from the viewpoint of the owner rather than an external position and to manipulate events or situations being investigated. Moreover, informal meetings with the participants were an important observation mechanism over the course of data collection. These meeting included 'catch-ups' over coffee, short discussions prior to the formal interviews and phone calls. Table 3-2 shows the events for which the researcher was invited to attend and participate. These events provided an opportunity to observe an overview of how IT and IT project management worked in the Council. Notes were made during each event, tidied up and reviewed after the event and then included in the data analysis.

TABLE 3-2: FORMAL AND INFORMAL MEETINGS ATTENDED

Type of Event	Duration	Topics discussed
Agenda for the ICT Portfolio Management Committee Meeting	2 hrs	• ICT portfolio report • Business cases and project management plans • Post-implementation reviews and reports
'Bold Future Vision' Presentation	1 hr	• A safe city where everyone belongs • The protection of the green environment • A clever design city by a strong and diverse economy • A community vision in 50 years time
iSPOT Training	All day	• Records and document management system • Tracking project history • Sharing feedback and reports
Informal Meetings	Over course of study	• ICT strategy and direction • IT projects, IT investments and public value • IT governance process, structures and relational mechanisms • Performance measures

3.5.2.2 ARCHIVAL RECORDS

Archival records are another form of data collection. It is a research approach concerned with the re-examination of existing recorded facts (Myers and Avison 2002). Public documents in electronic and hard copy formats were collected and reviewed. These included various government documents such as the IT governance vision, ICT strategy, reports, newspapers, the minutes of meetings, organisational records and mass media like government websites. These data are called secondary data and are an appropriate source for this research.

Other documents that were specific to the projects studied include concept plans, project charters, business cases, project management plans, value realisation plans and post-implementation reviews. These documents were collected and grouped into folders chronologically and according to the themes and projects, then uploaded to the NVivo 8 software for coding and analysis. Having data spanning long periods of time is essential, as they enable the researcher to look back in time, to cover large populations of people or other units of observations and to quantify reactions to events that the researcher cannot intentionally impose for practical or ethical reasons. This is not possible with other research methods (Hoyle et al. 2002).

Another important advantage of using historical documents is that a lot of information is gathered by governments and other organisations as part of their everyday operations, and it is often collected repeatedly. This helps to avoid the difficulties associated with people's awareness of being participants in research (reactivity) and often makes possible the analysis of trends over time. It provides the researcher with the opportunity to assess the impact of natural events and investigate many other issues including the appropriateness of using different types of methodology of IT implementation and the importance of engaging skilled staff. These documents have a high level of external validity because of the participants' unawareness of the research or its aims.

For additional information on the government's vision on IT projects, the following archival documents were also obtained through personal contacts and have been reviewed in terms of ICT strategy:
- ICT Governance in a Local Government (2008)
- ICT Governance in the Queensland government (2008)
- ICT Strategy 2005-2009 in a Local Government (2006)
- ICT Governance Framework for the Management of ICT Activities in a Local Government (2006)
- ICT Activity Assessment Model Version 0.01 in a Local Government (2006)

Since the above data was collected from sources of archival records such as government documents and was systematically analysed, the research adopted a content analysis approach. Content analysis is a widely used qualitative research technique. It is a formal technique for evaluating written or oral communications and analyses texts in the contexts of their users. It is a method of assigning quantitative value to words and phrases in order to assess the intensity and significance of a communication (Buckley, Buckley et al. 1976). Rather than being a single-dimensional approach, current applications of content analysis show three distinct approaches: conventional, directed or summative. All three approaches are used to interpret meaning from the content of text data and, hence, adhere to the naturalistic paradigm. The major differences among the approaches are coding schemes, origins of codes and threats to trustworthiness, all of which were addressed using the NVivo 8 software.

In conclusion, the analysis of archival records is advantageous when the researcher has to describe the incidence or prevalence of a phenomenon or when he/she is to be predictive about certain outcomes (Yin 2003). It is a technique for making inferences by systematically and objectively carrying out the researcher's analyses according to explicit rules that enable different investigators to obtain the same results from the same messages or documents (Holsti 1968, p. 324 as cited by Nachmias and Nachmias (1996)).

3.5.2.3 INTERVIEWS

One of the most important techniques of data collection is the interview method. It is an interaction involving the interviewer and the interviewee, the purpose of which is to obtain valid and reliable information. Interviews may take several forms: open-ended, structured and unstructured, focus group, in-depth, mail and phone. In our local government case study, semi-structured, open-ended and in-depth interviews with an email follow-up were conducted with 11 staff members from different levels of ICT governance departments.

In the semi-structured interview, the researcher participated with interviewees in their natural setting such as in their local coffee shop or office. A semi-structured interview was developed specifically as one of the instruments designed for this qualitative research. The data gathered by the researcher was observed firsthand in a natural environment, rather than being reported data. It was more flexible than a structured interview. It was a combination of a personal interview and reasonably extensive observation of actual social situations, thus providing a useful alternative to participant observation.

The unstructured interview method was useful for providing an oral history, where one or more participants were asked to recount aspects of their daily events, processes and interactions and the changes they have observed in the last five years (Yow 1994; Blaikie 2000).

In the open-ended interviews, the researcher asked key respondents for the facts of alignment maturity aspects as well as for their opinions concerning events in a particular project. Key informants were critical to the success of this research. They provided insight into certain occurrences and were a source of corroboratory evidence (Yin 1988). The open-ended interview questions were formulated to minimise research bias and provide an opportunity for respondents to openly reflect on their experiences, lending the interviews a purposeful meaning. Some questions were written in advance and were fairly fixed in terms of structure and interview direction while others were more flexible questions that allowed the interview to change and respond to the interviewee's responses.

Overall, the interviews were very advantageous. They were useful in obtaining a large amount of data within a reasonable time period. Using different informants allowed for a wide variety of information and a large number of subjects. There was also the opportunity for immediate follow-up questions by email for any necessary clarification. This in turn led to the in-depth interviews that allowed participants to respond with their views and experiences in their own words. Finally, conducting interviews gave the researcher greater control over the interviewing situation, as the respondents had to answer certain questions before they were asked subsequent questions. Biased responses, which arise when respondents provide too much information, were therefore minimised.

The above descriptions of the data collection were the actual survey instruments used in the proposed study. Interview questions were designed and counterbalanced throughout the different levels and course of study, based on the research questions and conceptual framework.

The interviewees were selected based on the conceptual research. Table 3-3 briefly outlines participants' job titles, roles and areas of responsibility as well as the interview duration. As indicated in Table 3-3, participants were selected from upper management and top executive levels since senior management have the greatest influence on the adoption and implementation of systems. The length of interviews ranged from 1.5 to 2 hours. The total duration of interviews was 673 minutes, equivalent to 11 hours and 22 minutes. The in-depth interviews were taken in the form of an individual interview (face-to-face), and data was gathered through written notes and audio recordings, which this researcher believes provides for more valid information. The physical place of all interviews occurred in three different locations in the local government office.

The researcher did not face any difficulty in finding participants (83%) to take place in the study, with an 83 per cent acceptance rate. This is largely due to assistance from the Chief Information Officer, who provided all the necessary materials and contacts as well as the place and time to conduct this research.

TABLE 3-3: PARTICIPANTS' PROFILES AND DURATION OF INTERVIEWS

Participants	Roles and areas of responsibility	Interview duration (mins)
Chief Information Officer (CIO)	• ICT strategy, IT governance • EA • Project management office • Organisational services	43
Executive Coordinator, ICT Portfolio Management	• IT governance • Office of ICT portfolio management • EA	53
Corporate Strategy Officer	• Developing organisation strategy properly such as quality control governance of corporate strategies	58
Executive Coordinator, ICT Strategy Policy	• Developing ICT strategy • Developing ICT policy • Developing EA	80
Manager, Corporate Planning and Performance	• Ensuring organisation delivery of a bold future vision • Ensuring that we have the right corporate governance frameworks and right plans • Controlling resource allocation, performance management and risk and internal audit	71
Project Manager	• Investigating infrastructure works of pipelines, water pumps station, etc.	59
Executive Coordinator of the ASAC Compliance Assimilation Data Specifier Section	• IT/business management team • Asset assimilation and inspected compliance team • IT projects team • Strategic delivery team	51
Executive Coordinator Asset Management, Strategy, Policy and Improvement	• Overlooking all asset management systems in the organisation and processes for ensuring data quality for those systems	52
Executive Coordinator Project Services	• Project services to manage project managers who are delivering ICT projects within the organisation	93
Project Management Officer	• Delivering projects and programs • Managing organisational change management' • Managing IT/business analysis	62
Benefits Manager	• Ensuring a robust benefits identification process • Embedding benefits process to business organisation	51

The interview questions covered a wide range of key issues to ensure that there was no relevant information missing which would threaten the internal validity of the study.

No matter how small our sample or what our interest, we have always tried to go into organizations with a well-defined focus-to collect specific kinds of data systematically (Mintzberg 1979, p.585).

These interviewees provided information that facilitated analysis of the data gathered during the case study. The analysis was also guided by an explanatory statement for participants (see Appendix D), executive summary of research purposes (see Appendix E) and a case study protocol (see Table3-4), which were created prior to data collection to enable the consistency and reliability of data (Yin 2003).

3.6 ETHICAL CONSIDERATIONS

The research study was designed to collect and triangulate data from several sources including the staff. In this process, problems of privacy, anonymity, confidentiality and other related issues tend to arise when the researcher tries to obtain valid and reliable data through participant observation and interviews. Each process in the research involves ethical consideration; hence, ethics concern was a matter of consideration in this study. The significance of ethics was made clear when the Bond University Human Research Ethics Committee (BUHREC) approved the study with a strict condition that participant identity restraint be maintained and no confidential public documents or individually identifiable narratives would be released to any third party.

Examples of ethical concerns in this research were the ethics protocol number, gatekeeper letter, informed consent, BUHREC approval and anonymity of subjects (Nachmias and Nachmias 1996) (see also Appendix C). The details of the research policy in relation to these ethical concerns are given below.

First, a BUHREC protocol number was obtained prior to data collection. It is a formal requirement by Bond University and an ethics application form will not be considered without this number. Then, the Explanatory Statement (ES) for research participants was given to the local government representative. This official letter provided information about research focus, voluntary participation, informed consent, anonymity issues and the nature of collected data. During data collection, the settings for research and all participants were highly respected and maintained so that participants were not put at risk nor was data disclosed for unintended purposes (Creswell 2009). Such a practice was followed to protect human rights in this research.

Next, the researcher obtained a gatekeeper letter from the local government. This letter outlined the mutual respect, benefits and trust between the organisation investigated and the researcher, therefore enhancing partnership collaboration between them and providing greater clarity in research direction and objectives. Informed consent was another practice implemented in this research. It was obtained through a letter with which explained that individuals could choose to participate in the investigation only after being informed of the facts that would be likely to influence their decision (Diener and Crandall 1979). The idea of this concept arose from both cultural values and legal considerations regarding the participants' freedom and self-determination. The consent form had to be signed by each participant, attesting their approval for interview participation. The form comprised such information as the purpose of research, confidential treatment of results, protection of participants and data under investigation, freedom to withdraw from research participation any time they wish and the avenue and duration of interviews.

The other methods used by the researcher to protect participants were anonymity and confidentiality, which were morally necessary to this research. It is unethical to violate research participants' anonymity and not keep research data confidential; hence, participant anonymity and data confidentiality needed to be ensured at all costs. Bond University has clear guidelines on the procedures concerning the storage of research data. The information provided towards this research will be kept in a locked filing cabinet for a period of five years from the completion of the research, after which it will be destroyed. The research participants were also provided with the details of the relevant people at Bond University who could be contacted should the participants have a concern or wish to complain about the manner in which the study was conducted.

In this research and after interviews were conducted, information was offered anonymously with the reader unable to associate a name with the data; the identity of the participant was then secured. Even if the sensitive information was revealed, it would not be able to be associated with an individual. In the semi-structured interview, anonymity was followed by linking names and other identifiers to the information by a code number. Moreover, the researcher will not publicly reveal a particular participant's information unless that information is subpoenaed by judicial authorities or legislative committees. Confidentiality is important in this research and can be enhanced by deletion of identifiers, crude report categories, micro-aggregation and error inoculation (Nachmias and Nachmias 1996).

In summary, ethical clearance for the study was obtained from the Bond University Ethics Committee, known as BUHREC, prior to the commencement of field visits and data collection. A gatekeeper letter was given to the researcher from the Council signifying approval of the local government's participation in this research. Each participant signed a consent form to approve his or her willingness to participate prior to the interview. Finally, the executive summary of research purposes was distributed to participants one week prior to collecting data, thus providing a general idea of the research focus.

3.7 BASIS OF QUALITATIVE RESEARCH DESIGN

The design of this research is based on research questions, the topic and the research methodology. Such a design reflects initial assumptions about how the study should be carried out, any legitimate problems that might arise, solutions and criteria of proof. Based on these considerations, this research adopted a qualitative approach. Golafshani (2003) describes qualitative research as a naturalistic approach that seeks to understand phenomena in context-specific settings, such as the "real world setting in which the researcher does not attempt to manipulate the phenomenon of interest" (Patton 2002, p.39) and only tries to unveil the ultimate truth. Strauss and Corbin (1990) state that qualitative research is any kind of research that produces findings not arrived at by means of statistical procedures or other means of quantification.

Unlike quantitative researchers who seek causal determination, prediction, and generalization of findings, qualitative researchers seek instead illumination, understanding, and extrapolation to similar situations (Hoepfl 1997).

To ensure the validity of the qualitative research design in this research, the following processes have been adopted, as suggested by Lincoln and Guba (1985), Hoepfl (1997), Patton (2002) and Yin (2003):

- The researcher uses a case study of a local government in its natural setting as the source of data. The researcher attempts to observe, describe and interpret settings as they are, where "quantitative measures cannot adequately describe or interpret a situation" (Hoepfl 1997, p.2). Thus, the methods of interviews and observations were adopted and were dominant in the naturalistic (interpretive) paradigm, consistent with positivist approach.
- The researcher acts as the 'human instrument' of data collection.
- The researcher uses inductive data.
- The researcher describes the characteristics of each setting, for example, the organisation, participants, IT projects, documents and research procedures, to gain a full picture of the case study undertaken, not only from the researcher's perspective but also from the reader's perspective. "If you want people to understand better than they otherwise might, provide them information in the form in which they usually experience it" (Lincoln and Guba 1985, p.120).
- The researcher uses a variety of empirical materials, case studies and documents that track the history of project process, interviews, interactional methods, personal interaction experiences, notes and live observations of participants in their natural settings, which aim to discover the meaning of the phenomenon undertaken and the meaning of events for the individuals who experience them.

Such methods describe an interpretive character, providing meaningful credibility to the research. Description is meant to convey or portray images of the case study undertaken, it also designed to persuade convince, express, or arouse passions. Descriptive words can carry overt or convert moral judgments(Strauss and Corbin 1990).

- Qualitative research is judged by its credibility and trustworthiness in its application and interpretation. Four paradigms were adopted to increase the validity of this research (see internal/external validity in Sections 3.7.2/3.7.3 in some detail).
- The researcher focuses on an emergent design or process as well as the outcomes or product of the research.

Validation of the research occurs throughout a research process, from the time of designing research questions to presenting the research analysis and findings. To understand the meaning of validity in qualitative research, it is necessary to present the various definitions of validity offered by many qualitative researchers from different perspectives. Four applicable paradigms are used as criteria for judging the quality of this qualitative research design. The paradigms are construct validity, internal validity, external validity and reliability, and are addressed in some detail in the following subsection.

3.7.1 CONSTRUCT VALIDITY

Construct validity is described as a way to establish a sufficiently operational set of measures for the concepts being studied so that objective judgments are used to collect the data (Yin 1988). Since construct validity is especially problematic in case study research, the researcher has to choose the specific types of changes that are to be studied and demonstrate that the selected measures of these changes do indeed reflect the specific types of change that have been chosen. According to Yin (1988), construct validity can be increased by using multiple sources of evidence, establishing a chain of evidence and having the key informants review the draft case study report.

3.7.2 INTERNAL VALIDITY (CREDIBILITY)

In this research, internal validity concerns raise the issue of 'whether we measure what we intend to measure'. In other words, it is the extent to which the study design enables it to accurately measure and study what it intends to study. That is, the researcher checks the accuracy of findings by employing certain procedures. These procedures, sometimes called measurements (Crocker and Algina 1986), are important to demonstrate that the query was accurately identified and described (Golafshani 2003).

Under such circumstances, it has always been difficult to completely ascertain that the variable which is being measured is the one for which the measurement procedure was designed (Nachmias and Nachmias 1996).For example, when a researcher is trying to determine whether event x led to event y without knowing that some third factor, z, may actually have caused event y. Identifying this type of error reduces the accuracy and consistency of the instrument used. Such an issue is called an internal validity threat, and its likelihood can be reduced by applying techniques such as pattern-matching, explanation-building and time-series analysis, particularly in this exploratory study, as supported by Yin (2003) and Winter (2000).

To strengthen the internal validity (credibility of findings) of this research, the following methods have been adopted. First, the triangulation method was used, as supported by Yin (2003), Patton (2002) and Mathison (1988).

Triangulation has risen an important methodological issue in naturalistic and qualitative approaches to evaluation [in order to] control bias and establishing valid propositions because traditional scientific techniques are incompatible with this alternate epistemology.   (Mathison 1988, p.13)

The triangulation method involves using multiple methods to collect sources of data. It is a useful strategy for increasing the validity of evaluation and research findings. Hence, it was used in in-depth interviews with different participants who have IT project and decision-making related backgrounds. In addition to interviews, a variety of methods (for example, analysing data and documents using NVivo), documents (for example, BC, PMP and PIR) and observation were used to increase substantiative evidence. However, the triangulation method does not suggest a fixed method for all researchers. The methods chosen in triangulation to test the validity and reliability of a study depend on the criterion of the research.

Further, causal relations and pattern-matching have been used within and across project analysis over time. Using an empirical pattern enhances internal validity when comparing it to existing literature or a predicted pattern. The method is convenient when examining all aspects of SA maturity (at different levels) and its impact on project outcomes are matched, leading to exploratory channel investigation to address research questions identified earlier.

Finally, both formal and informal revisits and feedback of predicted and discomforted findings and interpretations by actual participants, members of the participant organisation for verification and insight, supervisors and internal examiners were carried out throughout the research to sustain the credibility of trustworthiness (Rolfe 2006), defensibility (Johnson 1997) and confidence of the findings (Lincoln and Guba 1985). Discussions and feedback from different sources were a window towards the research's purposeful whole.

Together, all these verification strategies incrementally and interactively contribute to and build reliability and validity, thus ensuring rigor. Thus, the rigor of qualitative inquiry should be beyond question, beyond challenge, and provide pragmatic *scientific evidence* that *must* be integrated into our developing knowledge base.

(Morse, Barrett et al. 2002 as cited in Rolfe, 2006)

Thus, the abovementioned methods were followed to improve the overall credibility of this research, and to grasp a better understanding of the undertaken phenomenon. Nevertheless, as Professor Irava (2009, p.53) states:

From the current research status of the field and my issues of interest, using Patton's (2002) proposed typology, this dissertation is appropriately classified as basic research.

3.7.3 EXTERNAL VALIDITY (GENERALISABILITY)

Researchers use the concept of 'external validity' (generalisability) to refer to whether a study's findings can be generalised beyond the immediate study, that is, the external validity of applying results to new settings, people or samples (Yin 2003; Creswell 2009). However, generalisation in qualitative research is not the same as in quantitative research. In qualitative research, analysis relies on analytical generalisation, whereas in quantitative research, analysis relies on statistical generalisation. The researcher in analytical generalisation strives to generalise a particular set of results to some broader theory. A theory must be tested through replications of the findings.

In case studies, generalisability relies on rich, detailed descriptions of cases, participants and the study setting (Lincoln and Guba 1985). Thus, the question here is whether or not the case(s) described can be generalised to other settings and not to the whole population, as in the notion of the external validity in quantitative research. All considerations have been taken into account to improve the overall generalisability of this research.

3.7.4 RELIABILITY

Joppe (2000) defines reliability as the extent to which results are consistent over time. An accurate representation of the total population under study is referred to as reliability; if the results of a study can be reproduced under a similar methodology, then the research instrument is considered reliable. If the validity of trustworthiness can be maximised, then it leads to a more credible and defensible result (Johnson 1997). Reliability refers to examining the stability or consistency of responses. The most suitable terms in qualitative paradigms are credibility, neutrality or confirmability, consistency or dependability and applicability or transferability (Lincoln and Guba 1985).

Yin (2003) suggests documenting as many of the steps of the procedures as possible to determine if they are consistent or reliable. The reliability procedures have been identified by Gibbs (2008) as follows:

- Check transcripts to ensure that they do not contain obvious mistakes made during transcription;
- Make sure that there is no drift in the definition of codes and no shift in the meaning of the codes during the process of coding. This can be accomplished by constantly comparing data with the codes and by writing memos about the codes and their definitions;
- Coordinate communication among the coders by holding regular documented meetings and sharing the analysis; and
- Cross-check codes developed by different researchers by comparing results that are independently derived.

On the other hand, Patton (2002) puts forth three questions for the credibility (validity and reliability) of the qualitative research:

- What techniques and methods were used to ensure the integrity, validity and accuracy of the findings?
- What does the researcher bring to the study in terms of experience and qualification?
- What assumptions undergird the study?

The above suggestions were used as a guide to ensure consistent results in the research. I believe that the results of this research could be generalised to similar cases, settings and research methods, as described in this research (Johnson 1997; Stenbacka 2001) and "not the whole population as in the notion of external validity in quantitative research" (Irava 2009, p.71).

Table 3- 4 summarises the overall case study protocol of this research. Its function is to guide the researcher as well as participants to relevant SA issues in the undertaken investigation. Since qualitative researchers do in fact strive for reliability and validity in their findings, the case study protocol also serves as the procedure to be followed while writing reports on cases investigated, so that clarity will generate trustworthiness and meaning throughout the research.

TABLE 3-4: CASE STUDY PROTOCOL

Overview of case study project

1. Objective: Use empirical observations to refine, develop and provide valuable insights into a greater understanding of the perpetuation of IT project success in local public sector organisations.

2. Issue: The impact of SA perspectives on government IT projects

3. Research questions: See Section 2.7.

Field procedures

1. Access to site: Letter to organisation requesting permission to use public organisation site as case, explanatory statement indicating the nature and the scope of the study and consent form by participants.

2. Sources of information: Documents, website information, brochures, annual reports and meeting minutes, strategic plans, correspondence via emails and letters, interviews, etc.

3. Data collection plan: Scheduled interviews

4. Preparation prior to interviews: Ethics clearance form, forms requiring signatures such as gatekeeper letter, digital recorder and interview guide

Interview guide

1. SA perspectives

2. IT projects

3. Subject context (IT governance)

4. The meaning of projects success in organisation

Case study report outline

1. Case profiles
   a. Introduction/general
   b. Council profile
   c. Case summary characteristics

2. SA perspectives
   a. Strategy
   b. Knowledge
   c. Decision-making
   d. EA
   e. Public Value

3. Subject context: IT governance
   a. Structures
   b. Processes
   c. Relational mechanisms

4. Success of IT projects and maturity level of SA perspectives

5. Analysis, comments and conclusions

3.8 LIMITATIONS OF METHODOLOGY USED

Despite the efforts put in the research, in my bid to produce quality work, I still believe there is a room for improvements in methodology, which are discussed below. It is good to acknowledge these limitations and to understand how they can best be addressed during the research process so that shortcomings can be minimised.

Firstly, the most traditional criticism of a single case study is its inability to generalise to new settings, as each case has too many unique aspects (Blaikie 2000). However, the same criticism can be raised for a single experiment or the study of a single population, however scientists use those to generalise from one experiment to another (Yin 2003). A local government case study is a unique case and is conducted as an opportunity to observe and analyse a phenomenon that was previously less accessible to scientific investigation. It is worth documenting and analysing and therefore represents a significant contribution to knowledge for both practitioners and IT decision-makers. While empirical generalisation is affected by actual facts of the information gathered, the degree of generalisation in a research project is applied to express the knowledge in universal conformities law. In fewer cases, findings from qualitative research can sometimes be built on and used to base quantitative research studies later.

Secondly, characteristics of interview participants were limited to the local government where the case study was carried out. Most interviewees were at a high executive directorial level from the ICT governance section with high level of IT-related projects. While these interviewees are advantageous as purposive and reliable informants, the end-users at representative or operational levels would probably have better input than the top decision-makers. This problem was minimised by inviting these end-users to participate in the decision-making process in the IT governance committee (without being directly interviewed in this research).

Thirdly, it has been difficult to achieve reliability and validity due to the nature of qualitative data and availability of exploratory or subjective approaches in this research. In order to minimise this shortcoming, verification strategy was adopted. Verification is the process of checking, confirming and being certain. It is a mechanism used during the process of research to incrementally contribute towards ensuring reliability and validity (rigour) of the study.

Finally, direct observation and interview methods both introduce bias and influence the interviewer's perception and interviewees' responses. To remain objective and to avoid communicating personal views, other biases should also be taken into consideration such as unintended cues and nonverbal communication. This research reduced bias by building confidence between the interviewer and informants. The interviewer explained to all informants, in a friendly manner, the purpose of the study, its focus areas and the confidential nature of the interview. On understanding the significance and worth of the study, informants were cooperative with the researcher.

Despite all these limitations, I believe the research results have fulfilled the purpose of this study.

3.9 CONCLUSION

This chapter addressed qualitative research, in which a methodological approach was adopted to guide this research. Based on the research questions and nature of the undertaken study, reasons for selecting the research design and the chosen case study were provided. A local government in Australia provided an opportunity to better understand the impact of SA perspectives on IT projects in public organisations, and the findings could be applied and used to guide other public organisations. Units of analysis were adopted according to the research questions addressed, research context and research design. Hence, all the core units were included: IT governance components, SA perspectives and government IT projects. Various techniques were utilised to collect data: participation, observation in the real context of a local government, archival records and semi-structured interviews supplied rich information to understand the complexity of government IT projects. As with any research, relevant validity and ethical issues were also addressed. Limitations of the adopted methodology were briefly discussed, indicating further opportunities for using multi-case methodology in the future

4. A Case Study

4.1 General Remarks

Chapter 3 discusses the reasoning behind selecting the case study method as the chosen approach to explore the research questions. This chapter is concerned with the overall case study conducted at a local government and the data collected through the documents and interviews with various participants. The chapter is structured as follows. Section 4.2 provides the case analysis steps. Section 4.3 provides an overview of the case study brief. Strategic alignment issues cannot be understood in isolation and to better understand the concept of strategic alignment, it must be studied in the context from which it occurs. The section covers the s overview of the organisation, participant characteristics, case summary characteristics, characteristics of attributes and data sources. In other words, this section describes the procedure for gathering the data for the study. It also describes how the data and information was analysed and presented. Section 4.4 provides the meaning of success, followed by Section 4.5 which outlines the alignment maturity level. Section 4.6 provides as-is analysis of the case study (subject context). Section 4.7 summarises the content of the chapter.

4.2 Case Analysis Steps

As indicated in the discussion of case study protocol (refer to Table 3-4), analysis is organised and presented in three steps:

Step 1. The construction of the case profile.

A case profile was written up and a summary of the profile is presented.

Step 2. Coding for SA perspectives via the creation of nodes in NVivo 8. The SA perspectives nodes were classified into the Strategy, Knowledge, Decision-Making, Enterprise Architecture and Public Value Plan. Step 2 also explored ITG context (structure, process, and relational mechanisms) where perspectives occurred. Each of the individual contexts was also identified as ambiguous, negative or positive. The interpretation involved deploying these perspectives in a local government, this addressing Research Question 1 (RQa).

Step 3. Cross-project analysis will be highlighted to identify perspectives and attributes that have an impact positively or negatively on projects, thus addressing research question RQ(b) and RQ(c). The query tool in NVivo is also used to unearth those attributes that enhance on the success of IT projects.

4.3 Overview of the Case Study Brief

4.3.1 Overview of the Organisation

In the Australian federal system, there are three levels of government: Federal, State and Council (also called local government). Our case study is based on a local government with approximately 3000 employees. Its administrative government offices are smaller than a federal or a state government office. This research analyses 14 major local government IT projects. Local government bodies such as those which were used for this case study have specific responsibilities, duties and limitations concerning their areas of responsibility or influence.

4.3.2 PARTICIPANTS CHARACTERISTICS

The characteristics of the IT professional sample can be summarised as follows:
- They are the core source for ICT governance knowledge and practices;
- They have a high level of decision-making in IT investment and related IT activities;
- They are the owner and are responsible for clarifying the direction of business/IT strategy and their priority; and
- They use frameworks such as COBIT and ITIL to support their corporate strategy.

4.3.3 CASE SUMMARY CHARACTERISTICS

Table 4-1 shows a summary case of project characteristics. All projects were undertaken in the years between 2004 and 2009 and all were IT projects. The size of the projects varied from small (costing less than AU$500,000) to large projects (costing more than AU$2m) and had an implementation time period of less than six months to more than two years. The project documentation examined included Business Cases (BCs), Project Management Plans (PMPs) and Post-Implementation Reviews (PIRs). The projects were also classified into three types: well documented, less documented, and poorly documented. Their definitions are given below.
- Well documented: The project has three main documents (BC, PMP, and PIR)
- Less documented: The project has two main documents (either two of BC, PMP, PIR)
- Poorly documented: The project has one of any main document (one only of these BC, PMP, PIR)

TABLE 4-1: A SUMMARY OF CASE PROJECT CHARACTERISTICS

Projects	Size (AU$)	Duration	Documentation	Project Type
Project 1	under $500k	1-2 years	Less documented	IT-related project
Project 2	under $500k	6 months-1 year	Poorly documented	IT-related project
Project 3	under $500k	1-2 years	Less documented	IT-related project
Project 4	under $500k	1-2 years	Less documented	IT-related project
Project 5	under $500k	6 months-1 year	Less documented	IT-related project
Project 6	under $500k	6 months-1 year	Less documented	IT-related project
Project 7	$500k-$ 1m	Less than 6 months	Well documented	IT-related project
Project 8	$1m-$2m	6 months-1 year	Well documented	IT-related project
Project 9	$500k-$ 1m	1-2 years	Poorly documented	IT-related project
Project 10	$2+	+2 years	Well documented	IT-related project
Project 11	$2+	6 months-1 year	Well documented	IT-related project
Project 12	$1m-$2m	6 months-1 year	Well documented	IT-related project
Project 13	$1m-$2m	6 months-1 year	Less documented	IT-related project
Project 14	under $500k	1-2 years	Well documented	IT-related project

4.3.4 Characteristics of Attributes

The conceptual framework was developed based on literature and knowledge theory. Five perspectives of SA were identified. Each of these perspectives has three attributes. Their definitions are given below in Table 4-2.

TABLE 4-2: SA PERSPECTIVES AND THEIR ATTRIBUTES

Attributes	Characteristics
Strategy	
Clarity of Direction	Objectives are clear and communicated. Organisation has effectively communicated its vision and received understanding and 'buy in' from all employees and stakeholders.
Performance Measures	The flow of measurement mechanisms throughout the project, not necessarily to address a particular risk. These measures include the starting and finishing date, Critical Success Factors (CSFs), deliverables and milestones.
Quality of IT/ Business Plan	Accurate and enough details, how they affect other systems in the organisation, stakeholders, resources and estimated cost/benefits, vendors and SLAs are all addressed. A good plan has clear business and technical requirements.
Knowledge	
IT/Business Managers Participation in their Plan	Formal/informal participation between IT/Business managers to facilitate the implementation
Training	Needed training, workshops, tests, etc. for users as well as for staff across organisation. Sometimes a project depends on functional organisation
Decision-Making	
IT Investments and Budget	By whom decisions on IT investments and budget are made. It is a systematic approach of decision-making by who and how. Most decisions are effective and well communicated and based on accurate and detailed information on budget/benefits/risks estimation, and aligned with organisation strategy. The inputs of formal and informal decisions are given by committees, stakeholders, vendors, project management, team members and other staff. Some decisions can be made by individuals.
Prioritisation	A structural process is used to decide which IT investment is to be taken and at what cost. Reasons for selection of a project can be within a project itself or across other projects (programs). These reasons can be alternative options or risk status, criteria benefits, cost and alignment status
Stakeholders	Communication with all who have a legitimate interest. Key decisions are based on inputs from these stakeholders. They can be project team management, vendors, citizens, etc.
Enterprise Architecture	
Aligned Technical / Business Solutions	Flexibility of business/technical integration and following standardisation requirements. Flexibility of managing emerging technology.
Application and Technology	Architecture/Infrastructure (EA layers) are addressed in the process. For example, the technology priorities and policies allow applications, software, networks, hardware and data management to be integrated into cohesive platform. Enterprise network has no difficulty in integrating with other systems. Capacity or capability to link between what technology can do and business needs (problem).
Risk	What measures have been taken to mitigate risks? These measurements are specific to each project. However, they still follow other common measurements such as delivery of milestones, backup/ disaster, and capability to link to other systems without any problem.
Public Value	
Benefits to Organisation	Efficiency gains, operability, effectiveness, capability, operational level. Qualitative benefits, avoided costs and revenue generated
Benefits to Public	Improve customer services and customer satisfaction
Economic and Financial Metrics	Traditional metrics in economic and financial term such ROI, Net Profit, Bankable Financial Benefits

4.3.5 IT Project Characteristics

4.3.5.1 ICT Projects

ICT projects are those initiatives that are undertaken for implementation into the Information Communication and Technology (ICT) local government environment. ICT Projects enable the organisation to evolve its vision and deliver its strategy. The ICT project portfolio will be maintained by OS PMO and will be incorporated as part of a whole ICT portfolio View (ICT Activity Framework v. 15). Project characteristics have been identified by the PMBOK as the following:

- A project is a temporary endeavour undertaken to create a unique product, service or result. Temporary means that every project has a definite beginning and a definite end;
- A project creates unique deliverables, which are products, services or results;
- A key characteristic of a project is progressive elaboration which means developing in steps and continuing in increments;
- Project work differs from operational work primarily in that operations are ongoing and repetitive, while projects are temporary and unique; and
- Most of the projects have multiple stakeholders so they are impacted by more than business units.

A project is a temporary endeavour, having a defined beginning and end (usually constrained by date, but can be by funding or deliverables), undertaken to meet unique goals and objectives, usually to bring about beneficial change or added value (WIKIPEDIA 2011). Project management is known as a systematic approach to planning, scheduling, and controlling of a project and it is often managed by staff rather than tools (Kerzner 2009). The primary challenge of project management is to achieve all of the project goals and objectives while honouring the preconceived project constraints (Phillips 2003; Lewis 2006).

4.4 Defining Success

4.4.1 Defining the Meaning of Success

How do we define project success? Based on numerous perspectives on what is project 'success', providing a definition of 'success' will remain challenging, especially in a complex, changing project environment that involves a lot of stakeholders, such as is common in public organisations. In this case study, one participant argues whether it is a good idea to minimise variations in projects.

> Some strategic projects run over three years and if you were trying to minimise change in the project for the sake of it, you might miss a very valuable opportunity of, you know, recognising the change and doing something with it. So my personal belief it is not about minimising change within the project, it is about managing the changes. (Participant 4)

This view seems to be right where a project's scope changes because stakeholders agree additional value is possible through incremental increases in cost and time (Krigsman 2011). The degree of variation, however, should be reasonable. This view was expressed by one participant:

> But, I mean, there has to be some monitoring and we can't just let cost and time to exceed by 50 per cent for the sake of a benefit'. (Participant 8)

Hence, different staff in the organisation perceived the meaning of success differently. Such findings corroborate other findings in the research literature. In an article defining success, Ambler (2007)found that organisations have their own definition of success, and perhaps may even have different definitions of success for different types of projects.

While variation of project outcome is acceptable and encouragable due to the nature of projects, the rapid change of IT environment and to the benefits of organisation, other variation seems to not be acceptable because it occurred due to the lack of well-defined project plan and, importantly, because it reduced the benefits expected to accrue.

Well, when you look at that, what are the reasons why projects would end up being different from what you thought originally were? Part of the reason could be the estimation process in early on got wrong. Alright. That's almost certain to happen for us because we don't have a confirmed estimation methodology. There's no single one way the organisation handles all the details of planning of project, so you get variations'. (Participant 5)

Considering the many factors that contribute to project success, delivering on time and on budget is therefore no longer considered as an important measure of project success alone. Neither were checking off tasks and milestones on the project schedule considered enough. Milestones and scope were considered less important than outcome (a key indicator of project success) by one of participants:

I think when it comes to delivering milestones, project managers need some flexibility with regard to those milestones because, by nature, people move, they cannot make meetings. So from our prospective now is the focus is changing from focusing on milestones to focusing on the outcomes. So the true measure of the success of a project if the projects define the outcome. So you think about the outcomes - that is really what is important.

Saying scope is less important than outcome does not mean that scope can be ignored. Rather, two points should be considered. Firstly, the degree of tolerance is to be monitored during the life cycle of the project. Secondly, its clarity and direction ease the implementation and enhances the project success outcome. In fact, one of participants(Participant 7) states "the scope has to deliver to the quality required by the client to be considered a success." Another participant mentioned that:

..if the scope is unclear, the project will struggle and the benefit what is actually trying to achieve, the benefits is unclear and that's good luck for what you try achieving, I don't know, but it's hard to measure the success. (Participant 3)

When a participant was asked what the concept of success of IT projects means, the answer was:

Success in council means delivery of the benefits identified in the business case. It doesn't mean transformation of the business or delivery of true business outcome. And that to me is one of the key problems because the quantifiable benefits in the project are traditionally around technologies. Decommissioning of our technology replacement of that is not talking about the opportunities to change the way we do our business and deliver things better, quicker, smarter. Because they are too difficult to quantify so people don't put them in as benefits so the projects don't deliver them. (Participant 1)

Their answer calls for more speculation around the traditional interaction between business managers and IT managers. Unlike business managers, IT managers do not seem to go further than the technical outcome outlined in the business case to see what project can bring beneficial changes to organisation such as efficiency, quality and increased speed. They have different perspectives and, accordingly, interpret project success using their own way of understanding (Cleland and Ireland 2004). The implication of these findings indicates that it is vital for both business and IT managers to know what the project is trying to achieve after the completion time, therefore it is important that success criteria are clearly defined in the development stages.

The study by Standish Group in 2003, as cited by Ambler (2007), notes that 34 per cent of IT projects are successful, 51 per cent challenged (they are over schedule and/or they are overtime and/or they are missing significant functionality), and 15 per cent of projects are considered failures. If organisations, however, do not define project success in the same way that the Standish Group does, then the question is raised concerning what the actual success rates were. Hence, the key solution is to discuss in advance how to determine the success of the project. Another study by Ambler (2007) identifies five critical factors upon which IT projects are typically judged. These five factors are: schedule, budget, scope, quality and staff. Further investigation of these success criteria reveals the following:

- *Schedule:* 61.3 per cent of respondents said that it is more important to deliver a system when it is ready to be shipped than to deliver it on time;
- *Scope:* 87.3 per cent said that meeting the actual needs of stakeholders is more important than building

the system to specification;

- *Money:* 79.6 per cent said that providing the best return on investment (ROI) is more important than delivering a system under budget;
- *Quality:* 87.3 per cent said that delivering high quality is more important than delivering on time and on budget; and
- *Staff:* 75.8 per cent said that having a healthy workplace, both mentally and physically, is more important than delivering on time and on budget.

By far, it is critical for organisations to understand what the stakeholders consider as a successful project. Different perspectives of what success means are viewed differently by stakeholders. Consider the following statement:

Since I am acting as an 'xx'[15], I would say success means it met its objectives in its budgeting timeframe, that's what the project managers mean success…time and budget have met its outcomes, got those deliverables and made those outcomes. Now what Participant 9[16] would say because I run the business, success to me is that the business people have a working solution that gives them benefit and they are happy. (Participant 9)

Luckily, success criteria have changed considerably through time and moved from the traditional view of time and budget to a broader framework which includes value realisation and high-level user satisfaction.

There is a project plan in place that specifies costs at the low and high estimate. So as long as we deliver that project within those costs and we get the benefits then, I think, we are getting value for money. (Participant 8)

Having a clearly defined line to determine whether the project succeeds or fails seems to be a challenge for organisations, and continuous improvement is required, at least in current trends.

Well, usually with every business case or a project management plan and templates we have success factors and when you are doing the planning upfront you will actually say what the success factors are, so that's how you will measure success. We haven't been very good at them, obviously, because they are just things like the new system that will be implemented…we are actually going to measure before we start, but we haven't done that up till now…and once they get that focus of business improvement rather than an IT project, then we will start to get things like we will measure it first and measure it afterwards. That will be much, much better than staying in IT projects. (Participant 9)

Based on discussion above, the study reveals determining whether a project is successful or not is not nearly as easy as it might seem. Business drivers, different perspectives by stakeholders about anticipated benefits, the changing of focus on success factors through time as a result of the nature of the project and the rapid change of the technological environment or the nature of opportunity that precipitated the project all add to the challenges stakeholders and project management team face in defining unyielding project success. Taking all of these aspects into consideration, delivering a successful project starts with taking care that the project delivers the business results for which it was designed and which were clearly defined in its early stages. This can be achieved where all stakeholders, including clients, vendors and the management team, are working together as partners in identifying the reason that the project was launched, scoping the project and setting project goals and towards more specific criteria used to measure the benefits of a project. These are known as 'success criteria' or key performance indicators (KPIs). Therefore, it is important for organisations and clients to discuss how to determine the success of the project.

---

15   The actual name of a person was replaced by 'xx'

16   The actual name of participant was replaced by 'Participant 9'.

4.4.2 Defining and Identifying the Meaning of Success Factors and their Attributes' Value

Success factors are key indicators of project success. Scope, schedule, budget, risk, resources and quality are among the key indicators of project success mentioned in the PMBOK, 4th edition (Institute 2008). Another study by Ambler (2007) uses similar factors, except for 'staff' instead of 'risk'. Carter McNamara (2005) outlines seven possible definitions of what might be considered as 'success' in regards to projects. He, however, suggests that an organisation and client must work together to select one or more definition that best fits the organisation. These definitions are:

- The desired outcomes and results listed in the project agreement are achieved;
- The client's problem is solved;
- The project is finished on time and within budget;
- You and your client sustain a high-quality, working relationship;
- Your client learns to address similar problems by themselves in the future;
- Your client says that they would hire you again (if you are an external consultant); and
- You get paid in full.

Likewise, Crawford in his book *The Strategic Project Office*(2011), describes performance measures that link time, cost, and value to actual results. He broadens the success criteria specifically to address alignment between project goals, business requirements, and business outcomes. His list of performance measure is described as follows:

- The organisation's strategies are executed according to plan;
- The organisation's shareholders are satisfied;
- The organisation is financially successful;
- Projects are completed on schedule and on budget;
- Project customers are satisfied;
- Project resources are allocated optimally;
- Projects are aligned to the organisation's business strategy; and
- The organisation works on the right projects.

Obviously, success factors differ from one organisation to another. But, they have common inputs to upper management that lead directly or indirectly to the success of the project (Cooke-Davies 2002). Hence, the most important thing is defining success and failure in the context of business and organisation. In the case study in this book, four critical success factors are identified and supported by Ambler (2007); McNamara (2005); and ITGI(Institute 2008). These factors are budget, time, scope and value realisation (achieving benefits). Other factors related to operational levels such as staff and risk are excluded in the scope of this research. While they are good measurements to determine the possibility of failure or other negative situations, they are not evaluated in terms of the success or failure of a project. People and risk and their implications to project implementation are something that are going to be managed earlier not, at the completion of the project. Technically speaking and based on these definitions and the context of the case study organisation, it is possible to determine whether the project was successful or not according to the definition given below. The four success factors are defined as follows:

- **Scope (S)**: deliverables and milestones of the project. It is the foundation on which the schedule, budget and resource plans are built. Unless you get the scope right, the project will never be under control. The scope is defined by specifying an agreement on outcome. This will include assumptions, deliverables, functionality and technical structures. Hence, scope is important and deviation in scope can have an impact on time and budget and in turn reduce the benefits anticipated from the project.
- **Time (T)**: specifying the deadline of milestones and deliverables of a project from beginning to the end.

- **Budget (B)**: The Australian Macquarie Dictionary defines budget as an estimate, often itemised, of expected income and expenditure or operating results, for a given period in the future (Macquarie 2005). According to Investopedia dictionary (2011), a surplus budget means that profits are anticipated, while a balanced budget means that revenues are expected to equal expenses. A deficit budget means expenses will exceed revenues. Adjustments are made to budgets based on the goals of the budgeting organisation. In some cases, high variation of over budget can considered as a failed project. In our case study, a budget serves as functional resources allocated over the life of the schedule and deliverables of the project. It is, therefore, an issue and is one of the indicators or cause of failure. One project in our case study, Project 10, shows that budget was the reason that project was shut down, which means that the project is a failure, at least from a budget perspective.
- **Value Realisation (VR):** means anticipated benefits are achieved and occurred after the completion and run of a project. Such benefits known as the functional completeness of a project. It only occurs where many users go through their workflow in the new system (a project) several times.

These definitions are important. They are reference points which guide us toward a more complete understanding of the conditions that drive successful or failed projects. Simple definitions may not serve genuine business needs.

At this point, definitions of the project success factors were provided. The following paragraph identifies the attributes and the attributes value of these factors. All budget, time, scope, and value realisation factors have same four attributes, namely 'Achieved', 'Mostly achieved', 'Partially achieved', and 'Not achieved'. Their definitions and attributes value are provided as follows.

The meaning of criteria measurement and its equivalent value is depicted in Table 4-3.

TABLE 4-3: DEFINING CRITERIA MEASUREMENT

Attributes Measures	Definition
Achieved	Variation is equal to or more or less than 10% [$\pm^2$ 0≤10%] = 100%
Mostly achieved	Variation is plus or minus between 10% to 20% [± 10%≤20%] = 75%
Partially achieved	Variation is plus or minus more higher than 20% = 50%
Not achieved	Variation is very high, unknown, not applicable or are not measured = 25%

The meaning of success here is defined against four factors: Budget (B), Scope (S), Time (T) and Value Realisation (VR). These factors have been identified with their attribute values as Table 4-4shows. Attributes values (100%, 75%, et cetera) were chosen based on normal distribution (100% achieved, 75% mostly achieved) and the overall acceptance criteria used in organisation investigated. For example, when variation exceeds 20% (50%) of the original project plan, it is considered critical and must be resubmitted to Corporate Governance Committee (CGC[17]) or executive Leadership Team (ELT) for consideration and approval. 'Not achieved' criteria was not given 0%, but 25% to reflect the nature of projects which are not measured or abandon, fail to reach the final target, but some milestones, scope and other benefits are achieved or those projects have high variation.

---

17   CGC & ELT are endorsing authority

       '± stands for under or over budget'

TABLE 4-4: SUCCESS FACTORS AND THEIR ATTRIBUTE VALUE

Success Factors	Attributes	Attribute Value
Budget, Time, Scope and Value Realisation	Achieved	100%
	Mostly achieved	75%
	Partially achieved	50%
	Not achieved	25%

4.4.3 RANKING AND WEIGHTING (THE SIGNIFICANCE) OF SUCCESS FACTORS

As far as weighting is concerned, and based on interviews and documents, VR and B are ranked to weigh more than other two factors, T and S. Participant 8 clearly mentioned VR to be the most important:

> I think success is really time, budget and realising the benefits. Most importantly, realising the benefits. They are critical to the value proposition as to why you started the project in the first place'...So you think about the outcomes - that is really what is important. (Participant 8)

This view was also supported by another participant, who believes that T does not matter as far as VR and B are achieved.

> '..if it could be off-track in time, but not off-track on budget and not off-track from delivering the benefit, doesn't matter'. (Participant 3)

An additional follow-up survey was carried out with four participants to rank these factors from most important (1) to least important (4). The four participants were all working in a project management office and were relatively knowledgeable about project practices in the organisation. Table 4-5 shows the result.

TABLE 4-5: RANKING OF SUCCESS FACTORS

Interviewee	Budget	Scope	Time	Value Realisation
Participant 1	2	3	4	1
Participant 2	2	4	3	1
Participant 3	3	2	4	1
Participant 4	2	3	4	1

Table 4-5 shows that all four participants (100%) agree that VR is the most important factor, followed by Budget, which was ranked second-most important by three participants (75%). Likewise, three participants (75%) agreed that Time is the least important factor among all. Only two participants (50%) ranked scope as the third most important factor.

In defining the attribute values of success factors and ranking these factors according to their importance, the remaining step was to weigh the value of these success factors. We now know for instance, VR (or realising benefits) is the most important success factor followed by B. We, however, still do not know exactly how much percentage (weight) is counted for VR, for instance, and how much percentage is counted for B, and so on. Therefore, the above participants, including myself, were asked to weight the value of each success factor, considering the total percentage of four success factors is 100% and the ranking order obtained in Table 4-6.

In this way, it was assumed to follow the same rank order[18] obtained from the participants such as value realisation is the most important followed by Budget and Time is the least important. When there was

---

18   The second participant changed their rank order of Time success factor. This inconsistent response, however, has not affected the overall result obtained from all participants.

disagreement between participants, the average of weightage was taken and the result is presented in Table 4-5 below.

Table 4-6: Weighting of Success Factors

Interviewee	Budget	Scope	Time	Value Realisation	Total
Participant 1	25%	10%	15%	50%	100%
Participant 2	20%	10%	20%	50%	100%
Participant 3	30%	20%	5%	45%	100%
Participant 4	25%	15%	5%	55%	100%
Participant 5	25%	20%	5%	50%	100%
Average Weight Value	25%	15%	10%	50%	100%

These weight (shares) percentages should not be considered the constant or the sticky weights of each factor because they are subject to the nature of each project which may change their significance (value). However, the weights that have been derived seem plausible because those concerns were noted and addressed investigating each level within and across the attributes.

4.5 Designing the SA Maturity Level Measurement Instrument

The criteria instrument below shows the maturity level of SA perspectives and their attributes. The five maturity levels range between the lowest level of maturity (Ad Hoc) to the highest level (Optimised). Each of the perspectives above has attributes and their characteristics at various levels of maturity. For example, the 'Strategy' perspective has three attributes; 'Clarity of Direction', 'Performance Measures', and 'Quality of IT/Business Plan'. The detail of the attributes characteristics and their maturity is explained in Table 4-7 (See Also Table 6-1 and Appendix I).

Table 4-7: Maturity Level Criteria Adopted in this Study and Validated by Luftmann (2011)

	Attributes	Initial (1)	Committed (2)	Established (3)	Quantitatively Managed (4)	Optimised (5)
				Alignment Maturity Level		
Strategy	Clarity of Direction	Employees have very limited knowledge about the project and its purpose	Some goals are communicated	Most Objectives are clear	All objectives were clearly defined and well communicated	Effectively communicate its vision and get 'buy in' from all employees and stakeholders
	Performance Measures	No measurement mechanisms	Few measurements available	CSFs, milestones, deliverables and other measurement tools are established	Various measurements across levels and projects used to minimise risks	Measurement criteria are continuously improved to minimise risk
	Quality of IT/ Business Plan	Not clear, many questions have no answers	Some details of estimated cost, resources and benefits or ASLs	Details in cost, resources, benefits stakeholders and the impact to other systems	All milestones, deliverables, cost/ benefits, resources, stakeholders are will detailed, defined and accurate	All cost/benefits, resources, stakeholders, vendors, ASLs are well integrated and agreed up and aligned with organisation strategy

Knowledge	IT/Business Managers' Participation	Occasional participation	Few informal and formal participation	Participation is norm	Participation between them become culture in organisation	Their participation is in every single stage of project
	Organisation Emphasis on Knowledge	Knowledge is not shared with others. It is on the heads of people	Order and power focused with few shared responsibilities	Organisation engages all stakeholders and focus on sharing responsibilities	Information is spread equally across organisation and transparency is a key asset	-Variety of media is used to reach all stakeholders and employees - The culture of sharing power by knowledge is at the top strategy level
	Training	None	Minimum	Dependent on functional Organisation	At the functional organisation	Across the organisation
Decision-Making	IT Investments and Budgeting	Most decisions depend on individual people and risks/ benefits not addressed	-Few decisions depend on formal and informal structures - Low alignment with risks, benefits and organisation strategy - Cost focus	- Most decisions are made via formal and informal structures but some delay without action - Moderate alignment with risks, benefits, and organisation strategy	-Decisions are well supported and organised by formal structures - Well aligned with risks, benefits and organisation strategy - Key and minor decisions are dealt separately	- Inputs and endorsed formal and informal decisions are continuously improved - High and optimised risks, benefits and organisation strategy are all addressed
	Prioritisation	Selection process does not exist	Unstructured process exist and mostly introduced by business reference only	Have quantitative measurements such as options, risks and benefits criteria which IT investment is to be taken and at what cost	Structure process is well improved with options, risks status and criteria benefits	Prioritisation has a strong formal process
	Stakeholders	No decision inputs from stakeholders	Occasionally stakeholders give their inputs	Stakeholders are mostly engaged in decision-making process	-The inputs from stakeholders become a key and required aspect of organisation strategy	- Full engagement of stakeholders in all decisions-making processes. -Stakeholders are also informed of other decisions and planning
Enterprise Architecture	Align technical solutions with business solutions	Poor integration	Begun to adapt business/IT integration strategy	High business/ technical integration	- Standardisation requirements are specified - Highly business/ technical integration and flexibility	- Flexibility of integration - Business/IT solutions issues are all addressed
	Application and Technology	- Business/ organisation structures (the way the Council organises its businesses) are in initial stages architecture/ infrastructure domain does not exist - Ad hoc information and communication structures	- No flexibility in managing emerging network enterprise - A process of maintain application and managing IT infrastructure is in its beginning status	- Architecture/ infrastructure (EA layers) are addressed in the process - Enterprise network has no difficulty to integrate with other systems	- Strong adaptation of EA network - The technology priorities and policies, allow applications, software, networks, hardware and data management to be integrated into a cohesive platform	- Business/organisation structures are well matured - Architectures/ infrastructures structures are well mature - Information/ communication structures are well mature
	Risk Assessment	Risk not addressed	Few measurements used to mitigate risk such as deliverable and milestones	Variety of risk measures have been used at different levels including backup and disaster recovery	IT application and its capability to link to other systems without any problem	Effective risk measures are used through whole project

Public Value	Benefits to organisation	No benefits to organisation	Few benefits in plan such as efficiency gains and avoided costs	Efficiency, capability in operational level and other qualitative benefits are achieved	Most benefits are accounted such as efficiency, bankable benefits, effective operability and so on	-Efficiency gains, operationability, effectiveness, capability, operational level - Qualitative benefits, avoided costs, revenue generated
	Benefits to Public	No value to public	- Limited value to customers	Customer satisfaction focus	Customer services become a rule top priority to organisation	High quality of customer services
	Business/IT traditional financial metrics	No economic and financial metrics	- Limited focus on cost and efficiency metric	Some traditional matrix used such as ROI, Net Profit, Bankable Financial Benefits and Earned Value Management (EVM)	Shifting from only traditional matrix to other public value such as customer services	Using traditional matrix but much focus on Value Realisation matrix

4.6 As-is Analysis of the Case Study: Subject Context

The ICT Governance Framework establishes a system by which the current and future use of ICT resources is optimised. It includes the strategy and policies for using ICT within an organisation. The ICT strategies are decided by the Executive Leadership Team (ELT). The Corporate Governance Committee is responsible for endorsing ICT Projects that take into account ICT Strategy, the ICT Portfolio and resource requirements. The framework enables effective portfolio prioritisation and alignment with the local government's strategic objectives whilst aligning with Australian and Queensland Government ICT Standards.

The ICT Governance Framework includes the following points for management of the framework.

ICT Strategy: ICT Strategies are the strategic directions set by the ELT for managing the direction of the local government ICT environment. ICT Strategies must be aligned to the Corporate and Business Strategies and are seen as an enabler and support mechanisms for achieving corporate and business strategic goals.

ICT Principles: These are guiding principles that are applied when making decisions about the soundness of ICT initiatives and whether these initiatives should be supported and invested in. These principles assist in applying the Enterprise Architecture across the Council ICT Portfolio. An example of a guiding principle is to 'reuse before you buy'. ICT principles are used to guide investment in all aspects of ICT.

1. ***ICT Standards***: ICT standards are developed to ensure the ICT environment is stable and cost-effectively maintained. ICT standards are generally implemented and maintained through an Endorsed Products List. An example is to have a standard for an office productivity suite which enables collaboration in the workplace. The local government endorsed product is the Microsoft Office Suite 2000. At times, a product can meet multiple standards.

Enterprise Architecture: Enterprise Architecture is developed to provide an implementation roadmap for the ICT strategies. These roadmaps are developed to show how local government is going to transition its ICT environment to meet the Corporate requirements, and the merits of proceeding with these initiatives based upon the value and payback of implementation against the ICT strategy.

ICT Portfolio Management: ICT Portfolio Management is the ICT governance mechanism utilised by the business to manage the transformation of ICT from the current (as is) state to desired (to be) states through the effective coordination and prioritisation of all ACT activities, projects and investments.

The study also indicates that the interaction of structures, processes and relational mechanisms clearly brings greater explanatory power than a study of the SA perspectives of projects alone.

4.6.1 STRUCTURES

Every organisation has some structures, either specified or implied. Figure 4-1 shows the flow of decision-making mechanisms in local government. These committees and groups are responsible for monitoring the ICT activates across Council. They are described as follows: in the reference to the ICT Governance Framework of a local government, the Executive Leadership Team (ELT) is responsible for providing a clear direction of IT investment and determining the strategic business priorities while the Corporate Governance Committee (CGC) allocates the strategic resources that meet priorities set by the ELT. Achieving maximum business benefits and mitigating the risk of all individual corporate activities is included in responsibilities of this committee. This committee also assesses potential IT projects for their alignment with the local government strategic plan objectives (Council 2006).

Executive Leadership Team: The Executive Leadership Team (ELT) determines strategic directions and makes the decisions that affect the strategic direction of the Council. In reference to the ICT Governance Framework, the ELT will ultimately be responsible for approving the ICT Strategy. The ELT determines the strategic business priorities for business initiatives which guide the ICT Strategic direction for local government.

Corporate Governance Committee: The role of the Corporate Governance Committee (CGC) is to focus on strategic resource allocation for the organisation as well as prioritisation and monitoring of key corporate initiatives. The CGC ensures that the mission of individual corporate activities are aligned to strategic priorities set by the ELT, risk is mitigated, the achievement of business benefits is maximised and that initiatives are valued to ensure high value initiatives are given priority. The CGC is also responsible for ensuring that the all Information and Communication Technology (ICT) activities are endorsed by the Chief Executive Officer (CEO), taken into consideration with the ICT Portfolio and the necessary ICT resources are available to complete these ICT Activities.

Office of the Chief Information Officer: The Office of the Chief Information Officer (OCIO) is responsible for maintaining ICT governance processes and the strategies, policies and architecture within the Council. The ICT Governance Process will determine how the current and future use of ICT is directed and controlled. It involves evaluating and directing the plans for the use of ICT to support Council and monitoring this use to achieve plans.

The OCIO is responsible for providing the following support functions:

- ICT Portfolio Management;
- Enterprise Architecture;
- ICT Strategy and Policy;
- Capturing current and future strategic business ICT requirements;
- Coordinating the capture of new current and future strategic technology capabilities and providing analysis on the benefits to Council;
- Managing and maintaining the Business Solutions Reference Group; and
- Managing and maintaining the Technical Reference Group.

ICT Portfolio Management Committee: The ICT Portfolio Management Committee is responsible for the delivery of the Council ICT Portfolio. It is required to make recommendations regarding the ICT principles, standards and architecture. Additionally, it is responsibility for developing an ICT Portfolio Schedule that is to be used by the ELT and CGC when making decisions about Corporate project activity.

An ICT Portfolio Schedule is to be collated and provided to the CGC to assist the CGC in understanding the strategic ICT initiatives and their dependant ICT activities and also the resource and funding implications of each activity. The ICT PMC makes recommendations to the ELT and CGC as to the best methods for the delivery of all ICT activities that support the business initiatives.

The ICT Portfolio Management Committee is an ICT governance group that will collate the recommendations and direction from the ELT and CGC and provide ICT portfolio recommendations and scheduling advice on all ICT activities. Members of the group include representatives from City Governance, Office of the CIO, IT Operations Branch and the PMO.

The ICT Portfolio Management manages the local government's ICT activities by taking a holistic 'portfolio' view that drives the principles under which the management of the resources required for the portfolio of ICT activities is undertaken and supports the acceleration or decommissioning of activities. Portfolio management will take into account the business value within the portfolio of activities, alignment within the portfolio investment strategies, the likelihood of success in realising the benefits and the capacity of the organisation to deliver against the whole-of-life of the initiative. Recommendation against the ICT portfolio will be presented to the ELT or CGC for endorsement.

Business Solutions Reference Group: The Business Solutions Reference Group (BSRG) is an ICT governance group that provides advice and recommendations to the OCIO around business solutions to the organisations business processes and strategic and operational plans. Membership will be based on business domain representation (for example, assets, property, organisational management, et cetera). The approach to establishing the domains will be based upon current corporate priorities and current Council working groups.

The BSRG is an integral component of the governance structure of local government and is designed to enable business input into the development of ICT roadmaps through Enterprise Architecture. The BSRG function is primarily advisory in nature but the group's role can evolve with EA maturity.

The duties of the group include: making business solution recommendations; contributing to business solution subject matter expertise; providing business solution advice; and reporting on business solution risks and issues. The initial activities for the group will include:

- Establishing and maintaining the business solution principles and requirements for ICT in Council;
- Identifying and providing advice concerning the opportunities and risks of a business solution nature; and
- Assisting the OCIO ICT Strategy, Policy and Architecture Team with the conceptual development of business ICT solution roadmaps.

Technical Reference Group: The Technical Reference Group (TRG) is an ICT governance group that provides advice and recommendations to the OCIO around technologies that support business solutions and alignment to the organisation's business processes and strategic and operational plans. Membership to the TRG is based on the technology domains across local government (see Figure 4-1).

The TRG is an integral component of the governance structure of local government and is established to actively participate in and contribute to the governance of ICT through EA. The TRG is primarily advisory in nature but the group role can evolve with EA maturity.

Initial activities for the group include:

- Establishing and maintaining the technical principles and requirements for ICT in Council;
- Identifying and providing advice concerning opportunities, costs and risks of a technical nature; and
- Assisting the OCIO, ICT Strategy, Policy and Architecture Team on conceptual development of technical ICT solution roadmaps.

FIGURE 4-1: DECISION-MAKING STRUCTURES

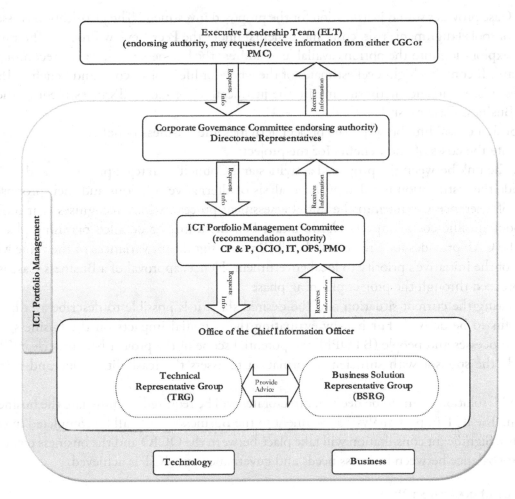

4.6.2 PROCESSES

The Corporate Activity Framework describes the process by which corporate activity is defined, identified and approved. The focus areas of Corporate Activity include Business Case, Project Management Plan (PMP), and the Post-Implementation Review (PIR). This researcher thoroughly examined these related documents on management practices to ensure high-value projects are given priority according to ICT strategic goals in this research. Hence, reviewing management practices and defining IT project management is needed (See also Section 4.6.2.2).

Such processes were examined over the time period of four years (2005-2009), which seems to better fit SA as process model as well as SA as a staged model as by definition the projects have set start and finish points, rather than shorter duration projects.

A concept plan is a brief one-to-two page description of the business objectives and the anticipated outcomes and benefits to the business, arising from a new idea or business need. It is the first stage of the initial business idea or plan for project management processes. The concept plan usually outlines the business need, stakeholders and different options for a solution.

Once the concept plan is registered by CPP, it is then submitted to CGC to determine whether it is a corporate activity or not. If it is considered under corporate activity, it is approved and the initial Project Peer Group needs to be identified and established prior to Business Case development. At this stage, no decisions are made about the project other than if it is a corporate activity or not.

4.6.2.1 BUSINESS CASE

The Business Case provides sound justification for the proposed investment. The justification is assessed based on a number of tools (alignment, risk, costs and benefits) under the Peer Review Process. The Business Case is the further exploration into the options available to address the business need, with a recommendation on the way forward. It contains high-level estimates of the whole-of-life project costs and benefits. The objective of the Business Case is to succinctly summarise the justification for Council's investment in the corporate activity. The Business Case must:

- Establish a causal link between the project activities and business benefits;
- Estimate the costs of and benefits for the project;
- Show the link between the proposed activities and Council's strategic priorities; and
- Provide the justification for change, an analysis of alternative solutions and their proposed scope.

A degree of tolerance is built into the initial assessment process, which recognises that much remains unknown about specific costs, impacts and resultant benefits until the detailed planning phase has been completed. The CAF provides for a review of the PMP where significant variances in the value will trigger a re-evaluation of the initiative's priority to local government. Hence, approval of a Business Case is limited to approval to proceed through the project planning phase.

By articulating the current situation and the desired state, it is possible to describe at a high level the business benefits to be derived. Further, by evaluating the potential impacts on the business, technology, organisation, processes and people (BTOPP), the potential scope of the project becomes clear. This exercise should provide the sponsor with sufficient information to assess the feasibility, value and priority of the project.

Where an IT solution seems to be needed, directorates will be required to enunciate the business need for an IT solution. The ICT impact analysis attachment to the Business Case shall be completed in conjunction with the OCIO. Significant consultation will take place between the OCIO and the business owners to ensure the appropriate balance between business needs and governance over ICT is achieved.

4.6.2.2 PROJECT MANAGEMENT PLAN

The Project Management Plan (PMP) is the implementation plan for the recommended solution with detailed costs and benefits as well as the specific timing and resource requirements. The PMP is the key reference for managing the project through to the delivery of business benefits. It records the principle elements of the investigation and planning research and summarises the deliverables for which the project manager, owner and sponsor are accountable (for example, costs, benefits, timeframes, et cetera)(Council 2007). The Project Management Plan must:

- Provide a full and complete cost/benefits analysis;
- Establish target measurable benefits for which the project sponsor is accountable;
- Contain base-line measurements against which progress toward benefits can be measured;
- Include a Benefits Realisation Schedule identifying how benefits will be measured and who will measure progress and timeframes;
- Include risk management and change management plans; and
- Provide the parameters for managing the project through to delivery of business benefits.

A methodology is a process designed for a set of methods and practices that are repeatedly carried out to deliver projects within a framework of intended objectives (Council 2007). The key concept is that the same steps are repeated for every project undertaken in order to gain efficiencies in the approach. By using a methodology you can:

- Provide a clear process for managing projects
- Create a project roadmap (charter)

- Monitor cost, scope, schedule and project quality plan
- Minimise risks and issues
- Manage staff and suppliers

Project Management Methodology is a systematic process used in Council for Information and Communication Technology (ICT) projects. It formalises the activities and responsibilities of the various parties involved in the ICT project initiation, planning, execution, controlling and closure. The use of this methodology will achieve a more coordinated and consistent approach to ICT project management across local government. This methodology is based on the Project Management Book of Knowledge (PMBOK), which provides a standard for managing projects. This standard is a collection of knowledge areas that are generally accepted as best practice in the industry.

It is important to select the elements of the methodology that are most suitable to each project undertaken. For instance, managing smaller projects requires lightweight processes when compared to managing large projects, which normally require monitoring and in-depth control of every element of the project. The methodology should tell the team what has to be completed in order to deliver the project, how it should be done, in which order and by when. A clear understanding of these elements will boost the chances of delivering a successful project.

The Project Management Methodology often exits in a broader context that includes:
- Portfolio management
- Program management
- Subprojects
- Project Management Office (PMO)
- Organisational Change Management Framework (OCM)
- Testing and procurement frameworks

Project management is the application of knowledge, skills, tools and techniques to project activities to meet project requirements. Project management is accomplished through the application and integration of project management processes. The project manager is the person responsible for accomplishing the project objectives.

Project managers often talk of a triple constraint: project scope, time and cost. Project quality is affected by balancing these three factors. High-quality projects deliver the required product, service or result with scope, on time and within budget. The relationship among these factors is such that if one of the three factors changes, at least one of the other factors is likely to be affected.

4.6.2.3 POST-IMPLEMENTATION REVIEW

The Post-Implementation Review (PIR) is performed on completion of the implementation phase of the project and is an overall assessment of the project's success. The purpose of the PIR is to address all project-related issues, both business and technical, including:
- Provide information to management regarding actual achievement versus planned achievement, with regard to cost and project deliverables;
- Review the Benefit Realisation Plan (BRP) and report benefit delivery progress;
- Identify actions to correct any outstanding issues;
- Provide an opportunity to learn from the positive and negative aspects of the project;
- Identify opportunities to add value to the CAF and project management processes; and
- Make any other recommendations on the future of the project and project deliverables.

The Review is not an audit. It is an overall assessment initiated by the Sponsor covering all aspects of the project.

4.6.3 RELATIONAL MECHANISMS

Relational mechanisms are guidelines used to manage the relationships between stakeholders, vendors, employees and customers. With clear mechanisms there will be always be a way to improve the outcome.

4.7 CONCLUSION

In this chapter, a case study was introduced in detail. This chapter established the research framework of the case study into three main sections. The first section discussed an overview of the case study brief. This includes in addition to the case analysis steps, characteristics and an overview of the organisation, participants, case, attributes and projects. The second section discussed the definition of 'success', identified the success factors of projects, their meaning, their attributes, their ranking and weightage level. The measurement of SA perspectives' maturity level was also introduced. The last section dealt with AS-IS analysis of the case study: research context. In this research context, fourteen projects were critically investigated and documented in Chapters5, 6 and 7.

5. ANALYSIS OF SA MATURITY WITHIN PROJECTS

5.1 GENERAL REMARKS

Chapter 5 discusses and analyses the SA maturity levels of IT projects and their success rates. This chapter consists of six sections. The first section (5.1) provides general remarks related to the chapter. Section 5.2 discusses the maturity levels of IT projects and then Section 5.3 summarises this maturity. In Section 5.4, a measurement of the success rates of these projects is established and then analysed within projects. Section 5.5analyses the relationship of SA maturity and success rates within project analysis and answers RQ(a), *'What is the relationship between the SA maturity level and the success rate of IT projects?'*. Section 5.6 further discusses individual success factors and their relationship to the maturity level. Hence, this chapter frames the main contribution of this book and the results obtained here are based on project analysis. The last section, Section 5.7, concludes the chapter.

The importance of within-case analysis is driven by one of the realities of case study research: a staggering volume of data(Eisenhardt 1989, p.540).

The volume of data generated by the research becomes more daunting because the search problem is often open-ended. For this reason, within-case analysis helps the researcher cope with this deluge of data.

5.2 PERSPECTIVE AND ATTRIBUTE MATURITIES: ANALYSIS WITHIN PROJECTS

This section depicts two dimensions of maturity: the perspectives and the attributes maturity levels of each project. The maturity levels were assigned as follows. Firstly, each attribute was defined in Table 4-7. Then, the rational description of the each perspective and attribute maturity level was described in Appendix I. The rationale description is based on the measurement instrument designed in Section 4.5. The maturity of all projects showed in this study is based on those steps.

Analysis of the individual maturities of attributes and perspectives provides a means of elucidating project improvement through time and, importantly, their degree of impact on project success. Identifying the characteristics of the maturities also helps clarify the conditions associated with a project (three attributes within each perspective) and whether they impact on project success or not.

5.2.1 PROJECT 1

Figure 5-1 shows a graphic summary of the maturity levels of all SA perspectives in Project 1. The maturity levels range from Ad Hoc (1) to Optimised (5).

FIGURE 5-1: ALIGNMENT MEASUREMENT IN PROJECT 1

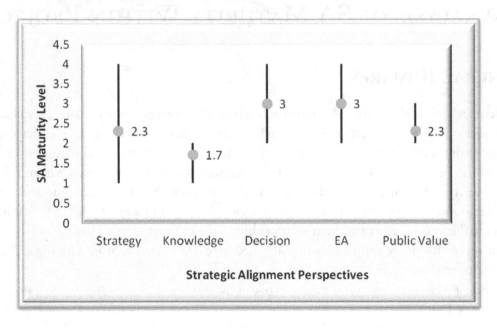

Perspectives maturity: The figure shows that Decision-Making (DM) and Enterprise Architecture (EA) perspectives receive the highest maturity level of Established (3), followed by Strategy (S) and Public Value (PV), both of which receive a Committed maturity level (2.3). The Knowledge perspective in this project, however, receives the lowest maturity level of Committed (1.7) when compared to other perspectives. This result may have several causes. While knowledge about the project was spread via the project management methodology and project reporting/responsibility hierarchy, this communication approach did not show a strong contribution to the knowledge of other staff who might be affected by the system. Hence, spreading the knowledge, sharing responsibility and planning were present but immature; possibly due to the fact that the Council had just formally adapted project management methodology and the process was still new to the organisation.

Attributes maturity: The figure also shows that the three perspectives, S, D and EA, receive an attribute of at least a healthy improved maturity level (4). However, S, again, and K also receive an attribute of at least an Ad Hoc maturity level (1), reflecting that the IT/Business plan, knowledge-sharing and performance measures were inadequate, at least in this project.

I don't think they have any good measures in place; most of it is anecdotal and emotional. (Participant 9)

Therefore, this project has perspectives maturity levels ranging from 3 to 1.7 and attribute maturity levels ranging from 4 to 1(See also Appendix I).

5.2.2 PROJECT 2

Figure 5-2 below shows a summary of the maturity levels of all SA perspectives in Project 2.

FIGURE 5-2: ALIGNMENT MEASUREMENT IN PROJECT 2

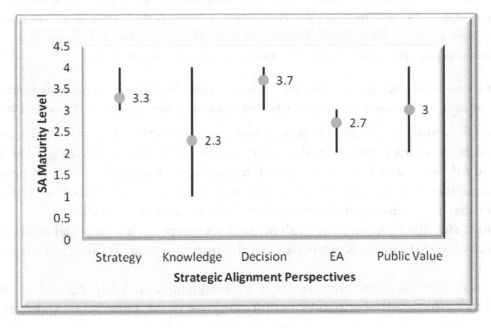

Perspectives maturity: The figure shows that D receives the highest maturity level of Improved (3.7), followed by S and PV, both of which receive an Established maturity level (3.3 and 3, respectively). The project shows the formal decisions on budget, which was the focus of many efforts. First, the budget expenditures, budget approval, quotes, all purchases, Purchase Requisition Numbers, project budget holders and cost estimates were all addressed by the team and project manager. A maturity achievement of 3.7 is the highest maturity reached by any of the projects investigated in this study.

This perhaps reflects the importance of the budget in the eyes of the Council:

That's getting better. When you look back at the budget process last year for the first time, it wasn't altered by ELT and wasn't altered by Council. So that's actually symptomatic good process because they recognise the business actually supports the IT budget because they saw their priority reflected on it and because of that the council didn't rip it apart because the business went as a united unit to council and said this is what we actually need. So that's good... So at the strategic level... (Participant 5)

Second, engaging stakeholders in different IT activities throughout the project played a vital role in decision-making. Key decisions in Project 2 were made by groups and not by individuals within the Council. For example, senior/executive management used the Project Charter to approve requests for all resources required. Additionally, measurements of the project's value and other decision criteria were used to select and prioritise the project.

So, I mean, part of that role was, well, obviously we had to make very good relationships with people across council because it is easy then to be doing, but it's about building that relationship so that's - essentially they understand that we are not going to accept it, not understand it, so eventually you get to the point where they are producing better material. (Participant 2)

The Knowledge perspective, on the other hand, has the lowest maturity level (Committed, 2.3) compared to other perspectives in this project, but shows an improvement compared to Project 1 (1.7). Interestingly, the Council used different forms of communication to emphasise knowledge; for example, when the Council explained why the system was important and needed to be implemented. All correspondence was done via email, Microsoft Word, Microsoft Excel, Visio or PDF documents and meeting minutes. All matters relating to Project Delivery were communicated to the project manager via email or through project meeting minutes.

Issues, risks and actions resulting from these communications were documented and tracked by the project manager. A detailed communication plan, included in the implementation management plan, complied with the vendor's required activity:

We have got some very good mechanisms in place for communication now. We have got good management structure; we are following a life cycle; we've got a corporate activity framework where we go through Gateways; we consult across the organisations; we have a number of committees and focus groups and forums that we attend and we discuss the different initiatives as they come up; we've got corporate portfolios in place; we've got corporate reporting in place, the corporate governance committee in place. So I think we communicate reasonably well. (Participant 4)

So I think we are starting to move quite quickly in that way now. But there has been a definite improvement in communication and in knowledge-sharing and in consultation and the relationship, it is really quite good now. (Participant 4)

However, the Knowledge perspective in general was affected by other attributes, namely communication between IT and business managers, as well as training, with both factors receiving a low maturity level (Committed and Ad Hoc, 2 and 1, respectively). While training the users and the support personnel and IT/business communication had been mentioned among the strategic goals, in practice no formal training was provided. Other than the Training attribute, all other attributes are at the middle level.

Attributes maturity: The figure also shows all perspectives except EA have received an attribute of at least improved maturity level (4) and the K perspective has the lowest attribute, Training, which received Ad Hoc maturity level (1).

In short, this project has perspectives' maturity levels ranging from 3.7 to 2.3 and attributes maturity ranging from 4 to 1. The details of attributes maturity are given in the Table 5-2 in Appendix I.

5.2.3 Project 3

Figure 5-3 shows a summary of the maturity levels of all SA perspectives in Project 3.

Figure 5-3: Alignment Measurement in Project 3

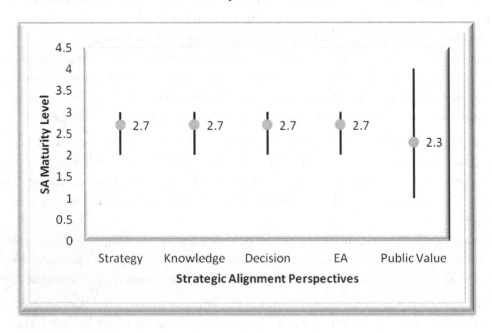

Perspectives maturity: The figure shows that all perspectives have an equal maturity level of Established (2.7), except Public Value (PV), which has a Committed (2.3) maturity level. This project shows a bit of improvement in K perspective compared to the previous two projects.

Attribute maturity: The figure also shows that all perspectives have attributes maturity levels between

Established (3) to Committed (2), except PV, which has attributes maturity levels between Improved (4) and Ad Hoc (1), indicating that the overall average maturity level of this project is not high. The major limitations in this project's maturity relate to missing key information in the business/IT plan in both BC and PMP and the project itself was considered as less documented (see Table 4-1). There was no PMP for Phase 1 and 3 and no PIR for all three phases. Missing key information can cause failure of the project, as indicated by one of the participants:

> But why is the project burning money? A lot of times because they haven't got a well-defined scope, well and clear benefits… So you go look at the causes of burn rate. There is a cause in burn rate if things aren't well understood.

(Participant 3)

5.2.4 PROJECT 4

Figure 5-4 shows a summary of the maturity levels of all SA perspectives in Project 4.

FIGURE 5-4: ALIGNMENT MEASUREMENT IN PROJECT 4

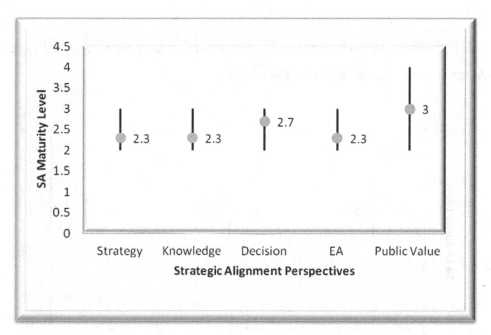

Perspectives maturity: The figure shows that the PV perspective receives the highest maturity level of Established (3), followed by the DM perspective, which receives a slightly lower Established maturity level (2.7). The high maturity of these two perspectives was perhaps due to the fact that service continuity was given high strategic priority and the key stakeholders' plan and impacted directorates were involved formally in decision-making, in accordance with Business Continuity Plans 2005-2006 and with the Corporate Plan and Corporate Strategic Priorities. The concept of public value in this organisation was given by two participants as follows:

> Within benefit realisation templates, which we encourage people to use to define the project plan. Out of our template we have identified seven potential benefits. Four are non-financial benefits: strategic priority, delivery, risk mitigation, customer service improvement. The efficiency gain people can do the same with less: for example, customer system. One of the key benefits of a telephony system is to allow existing staff to deal with an increased amount of enquiries to deal with increase in population; so if we didn't have a better system that would not be possible, so it's doing more with your existing labour. If we have a different system, that would not be possible; that is classified as an efficiency gain in this organisation. (Participant 8)

The governance makes sure you are investing in the right thing, so you provide money to a project to go and deliver. There is another governance function to make sure it's delivering what you pay for it to do. (Participant 3)

The findings also indicate that the maturity of S decreases in this project (2.3). Moreover, three perspectives (Strategy, Knowledge and EA) have low levels of maturity, Committed (2.3). This low maturity is due to a combination of factors in these perspectives, including deliverables, key stakeholders, the lack of provision of external resources and the rare communication between IT/business managers. Additionally, the only plan for a rapid resumption of Council services offered solutions to mitigate a variety of risks, such as flood, bushfire, earthquake and disease.

Attributes maturity: The graphic also shows that all perspectives have attributes maturity levels between Established (3) to Committed (2), except PV, which has attributes maturity levels between Improved (4) and Committed (2).

In short, the project has perspectives maturity levels ranging from 3 to 2.3 and attributes maturity levels ranging from 4 to 2. As indicated in Figure 5-4, this project does not demonstrate any Ad Hoc maturity level (1).

5.2.5 PROJECT 5

Figure 5-5 below shows a summary of the maturity levels of all SA perspectives in Project 5.

FIGURE 5-5: ALIGNMENT MEASUREMENT IN PROJECT 5

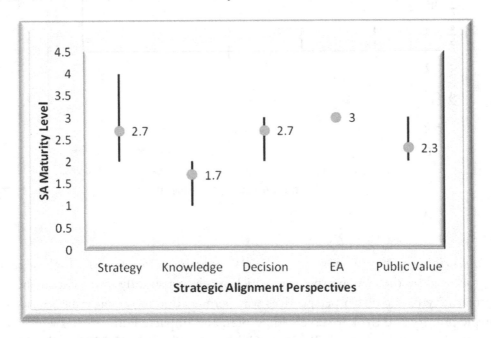

Perspectives maturity: The figure shows that EA receives the highest average maturity level of Established (3) and the Knowledge perspective receives the lowest average maturity level, Committed (1.7). Sharing responsibilities in the transaction plan, accountability and the relationship with the Certifying Body were all broadly addressed. Based on this research, effective process management from concept plan to post-implementation review in this project and training are perhaps the areas most in need of improvement by the Council:

So my focus within the PMO has been on maturing up a lot of the existing processes so that we can manage these large programs to work very efficiently and effectively, but also that we have, like, a consistent, repeatable defined process for everything so that everything is done in the same sort of ways across programs and projects, not moving away from them, each being slightly different in a way, so even though they are consistent repeatable processes, there is an element of ad hoc in that, yes. (Participant 11)

Attributes maturity: The figure also shows that unlike other four perspectives, the K perspective has at least one attribute receiving Ad Hoc maturity level. 'Clarity of Direction' receives the highest maturity level of Improved (4) while the attribute 'Training' is the only attribute receiving the lowest maturity level of Ad Hoc (1).

In short, this project has perspectives maturity levels ranging from 3 to 1.7 and attributes maturity levels ranging from 4 to 1.

5.2.6 Project 6

Figure 5-6 shows a summary of the maturity levels of all SA perspectives in Project 6.

Figure 5-6: Alignment Measurement in Project 6

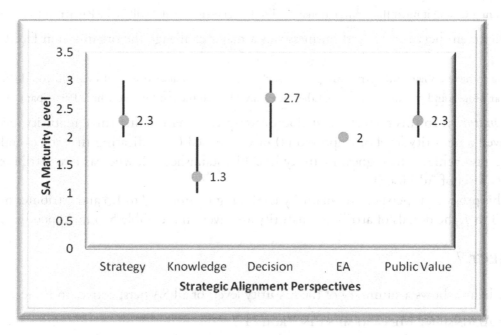

Perspectives maturity: The figure shows that the Decision-Making perspective receives the highest average maturity level, Established (2.7), whereas the Knowledge perspective receives the lowest average maturity level, Ad Hoc (1.3). Such maturity (1.3) is the lowest maturity received across projects. The main challenge facing this project was again identified as the lack of understanding, particularly between business and IT managers:

Well, I think the actual business managers of council need to understand what technology could do for them in terms of benefit rather than in terms of technology. Here is the technology... [which] is looked at. It really needs to apply technology. So they almost need to apply to an education program where they are saying this technology could do these kinds of things for you, so you get an environment where business is actually partner and actively wants IT to come in. Whereas currently that's not the case... it is IT. (Participant 1)

Participant 11 responded as follows to the question, 'What challenges did you face regarding IT projects?'

Knowledge, good understanding of business/IT planning. I think in the knowledge we would be pretty much a three. There are set workshops and things for planning out the ICT Agenda, although I believe their focus is slightly wrong, but like I said before, that we are where we are because we are. And Strategy, I would probably say we are closer to a two but again between a two and a three. (Participant 11)

When the question 'So what do you think is the role of the impact of knowledge for the success of IT projects?' was asked of another participant, the answer was:

Huge and we do it so badly. (Participant 4)

In answering the question, 'In what ways can the council improve business and IT communication?', one staff member answered that the organisation uses every possible opportunity to improve the situation:

I think there are few methods are currently used; a newsletter for big projects, regular communication via emails, providing updates and progress on projects to all staff. So that, I guess, is to get buy-in to the change. (Participant 8)

Other perspectives (Strategy, EA and PV) have the average maturity level of around Committed (2). It should be noted, however, that the maturity level of the EA perspective (2) in this project is considered the lowest maturity received among the EA perspectives of all projects examined in this study. The overall maturity level of EA perspective in organisation was predicted as Committed (2) by one of the participants:

Enterprise architecture, they've certainly got it so it's a, erm... that's tricky because 3 said integrated across the organisation. Now it's not, so I'm guessing it must [be] transactional level so I'm guessing level 2 will be it.(Participant 7)

Lack of alignment between IT and business was a major challenge the organisation faced in designing its EA:

We are going to have a major infrastructure project such as a new parklands or something and you don't know, that's a two- or three-year project and you don't know that they need any IT until the last two months. (Participant 3)

Attributes maturity: In this project, all attributes are spread over the first three maturity levels (1–3). No attribute received a maturity level of Improved (4) or Optimised (5), indicating that the overall maturity is low. Four attributes achieve the highest maturity level of Established (3), whereas two attributes receive the lowest maturity level of Ad Hoc (1).

In short, this project has perspectives maturity levels ranging from 2.7 to 1.3 and attributes maturity levels ranging from 3 to 1.The details of attributes maturity are given in the Table 5-6 in Appendix I.

5.2.7 PROJECT 7

Figure 5-7 below shows a summary of the maturity levels of all SA perspectives in Project 7.

FIGURE 5-7: ALIGNMENT MEASUREMENT IN PROJECT 7

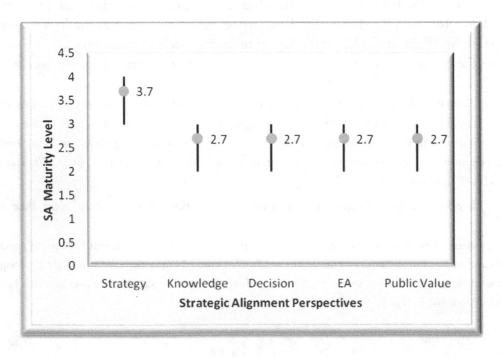

Perspective maturity: Interestingly, all perspectives except Strategy receive an average maturity level, Established (2.7). The Strategy perspective receives the highest average maturity level, Improved (3.7). This high maturity of Strategy indicates that the project's IT/business plan, direction and performance measures were all clearly identified, explained and communicated in detail. Amongst all maturity levels in the S perspective across all projects, this project receives the second highest maturity level (3.7), after the maturity level of Improved (4) of Project 8.

Attributes maturity: In this project, two attributes, Clarity of Direction and Performance Measures received the highest maturity level, Improved (4). Unsurprisingly, both high-level attributes come from S perspectives. Strategy perspective has been and will remain a core direction and vision of the organisation. The remaining attributes spread between levels 3 and 2.

We have got a 30-year vision, we have got a five-year corporate plan, we have got a range of corporate strategies which are delivering on those, so we have got quite a lot of mechanisms in place to make sure that the programs and projects we are delivering, are delivering the right things … so we have now got a lot more strength in our current corporate strategy, so a lot of our outcomes have already been identified, we know what they are for our corporate plan for the next five years in the bold future vision. So when we are developing strategies now, whether it's ICT or other things, they are really not developed in a vacuum; they are being developed and so really the outcomes of those are already known.(Participant 10)

The figure also indicates that the project does not receive an Ad Hoc maturity level (1) for any of its attributes, indicating the project's overall maturity is high.

In short, this project has a perspectives maturity level ranging from 3.7 to 2.7 and an attributes maturity level ranging from 4 to 2. The details of attributes maturity are given in Table 5-7in Appendix I.

5.2.8 PROJECT 8

Figure 5-8 below shows a summary of the maturity levels of all SA perspectives in Project 8.

FIGURE 5-8: ALIGNMENT MEASUREMENT IN PROJECT 8

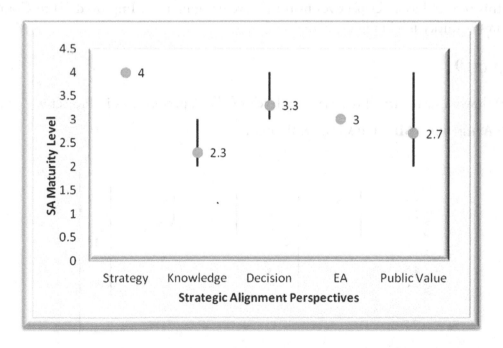

Perspectives maturity: The figure shows that the Strategy perspective receives the highest average maturity level, Improved (4), while the Knowledge perspective receives the lowest average maturity level, Committed (2.3):

I just think there's lack of business/IT understanding. IT is only here for business and I just think there's a poor knowledge, certainly in ICT area and the business area. The business is often not very clear what it does as well. So there's two people there. (Participant 7)

The high maturity of Strategy is due to the presence of accurate and sufficient details about highly technical requirements, the quality of the IT/business plan, measurements and objectives, all of which were clearly defined and communicated.

Attributes maturity: In this project, five attributes receive the highest maturity level of Improved (4), whereas four attributes receive the lowest maturity level of Committed (2). The remaining six attributes were spread across the middle, around level 3. Overall, the project shows its improvement maturity level in all attributes except in two attributes in the Knowledge perspective. The Knowledge perspective remains challenging for many projects and needs to be addressed. This frustration can be heard from the top IT management team:

It really depends on the project itself. And I run a lot of different projects as well; it can be upgrading systems and introducing new systems, so different challenges for different projects, but some of the challenges that we face that are common to most of the IT projects that we do is we are always challenged by finding an appropriately skilled and experienced project manager. What totally amazes me is this we are in the construction business… we buy in contractors, what don't contractors know - they don't know our business, they know project management, they do not know our business, they know methodology, life cycle, et cetera, et cetera, and that is one of the biggest risks to projects. I think a methodology is a guide it is only as good as the person that is using it.

(Participant 4)

The figure also indicates that both Decision-Making and Public Value perspectives have at least one attribute receiving an improved maturity level (4). In addition, both Knowledge and EA perspectives have at least one attribute receiving an Established maturity level (3). This project does not demonstrate the lowest maturity level of Ad Hoc (1) in any of its attributes. In fact, the lowest maturity level of attribute in the S perspective is Improved (4).

In short, this project has a perspectives maturity level ranging from Improved (4) to Committed (2.3) and an attributes maturity level ranging from 4 to 2.

5.2.9 PROJECT 9

Figure 5-9 shows a summary of the maturity levels of all SA perspectives in Project 9.

FIGURE 5-9: ALIGNMENT MEASUREMENT IN PROJECT 9

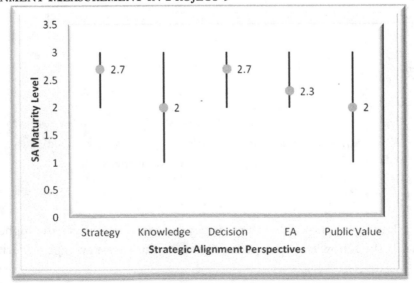

Perspectives maturity: The figure shows that the Strategy and Decision perspectives received the highest average maturity level of Established (2.7), while Knowledge, EA and PV received around a Committed maturity level (2). As in Project 2, this project does not receive a perspective maturity level above 3, indicating the challenges it faces in overall maturity. In this project, the PV perspective receives the lowest maturity level (Committed, 2) of any PV perspective in projects examined in this study. Demonstrating the value technology brings to the business goals is not new and has been a challenge in earlier projects:

I think the difficulty in the past has been because we are very much seeing the projects as ICT projects; the value to the business in the community hasn't been evident. (Participant 10)

Attributes maturity: In this project, most attributes were spread over levels 2 and 3. The figure also indicates that, unlike Project 8, this project does not receive an Improved (4) maturity level in any of its attributes. In addition, the project has at least two attributes with a low maturity level (Ad Hoc, 1). These attributes were Training in the Knowledge perspective and Economic/Financial metrics in the Public Value perspective.

In short, this project has perspectives maturity levels ranging from Established (2.7) to Committed (2) and attributes maturity levels ranging from 3 to 1.

5.2.10 PROJECT 10

Figure 5-10 below shows a summary of the maturity levels of all SA perspectives in Project 10.

FIGURE 5-10: ALIGNMENT MEASUREMENT IN PROJECT 10

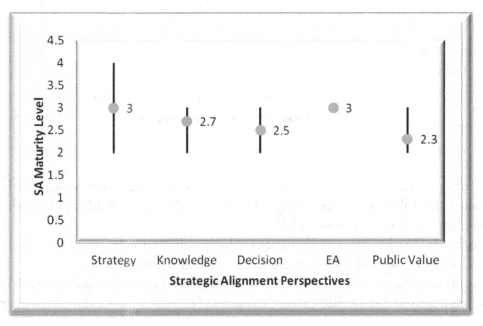

Perspectives maturity: The figure shows that the Strategy and EA perspectives receive, on average, the highest maturity level of Established (3), while the PV perspective receives the lowest average maturity level, Committed (2.3). Note that the previous four projects receive the same maturity level of Committed (2.3) in their PV perspectives, indicating that the PV maturity has not improved much over time.

Attributes maturity: In this project, the Clarity of Direction attribute receives the highest maturity level, Improved (4). All other attributes are spread over levels 2 and 3. Therefore, none of any attributes receives a maturity level of Ad Hoc (1) or Optimised (5). Far from maturity, this project faced many challenges and was abandoned before its final implementation. Some of these challenges are explained by PIR internal auditors (Council 2009, p.8)as "little is known about the scope planning work undertaken for this project".

Where the difficulty arises is in the articulation and identification of those business benefits identified in the BC and PMP documents, and more importantly, the measurement and evaluation of those business benefits. Since ICT systems that don't always have bankable benefits associated to them, a lot of the value of ICT isn't tangible.

In short, this project has perspectives maturity levels ranging from Established (3) to Committed (2.3) and attributes maturity levels ranging from 4 to 2.

5.2.11 PROJECT 11

Figure 5-11 shows a summary of the maturity levels of all SA perspectives in Project 11.

FIGURE 5-11: ALIGNMENT MEASUREMENT IN PROJECT 11

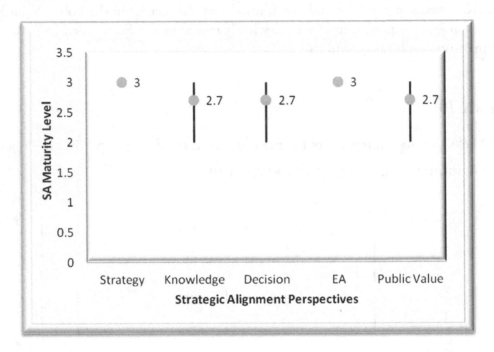

Perspectives maturity: The figure shows that both Strategy and EA receive, on average, the highest maturity level of Established (3), followed by Knowledge, Decision-Making and Public Value, which receives the slightly lower maturity level of Established (2.7).

Attributes maturity: In this project, all attributes have only two maturity levels, Established (3) and Committed (2). Therefore, none of the attributes receive maturity levels of 1, 4 or 5.

In short, this project has perspectives maturity level range around Established (3 and 2.7) and attributes maturity levels ranging from Established (3) to Committed (2).

5.2.12 PROJECT 12

Figure 5-12 below shows a summary of the maturity levels of all SA perspectives in Project 12.

FIGURE 5-12: ALIGNMENT MEASUREMENT IN PROJECT 12

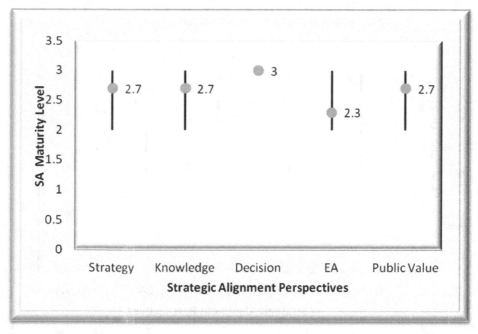

Perspectives maturity: The figure shows that the Decision-Making perspective received the highest average maturity level of Established (3), followed by the Strategy, Knowledge and Public Value perspectives, all of which received a maturity level of Established (2.7), while the EA perspective received the lowest average maturity level of Committed (2.3). The high level of the Decision-Making perspective is due to the shared responsibilities in decision-making regarding IT investments, internal resources, cost/benefits and alignment with economic visibility and organisational priorities. Thus, stakeholders were highly engaged in decision-making processes.

Attributes maturity: In this project, as in Project 11, all attributes are located in only two maturity levels of Established (3) and Committed (2). Therefore, none of the attributes receive maturity levels of 1, 4, or 5. In addition, two attributes, Organisation Emphasis on Knowledge and Training, both receive a maturity level of Established (3), indicating the improvement of the Knowledge perspective from previous projects.

The figure also shows that the Decision-Making perspective received no attribute maturity level below Established (3), while the other perspectives had attribute(s) at least one lower maturity level of Committed (2).

In short, this project has perspectives maturity levels ranging from Established (3) to Committed (2.3) and attributes maturity levels ranging from Established (3) to Committed (2).

5.2.13 PROJECT 13

Figure 5-13 shows a summary of the maturity level of all SA perspectives in Project 13.

FIGURE 5-13: ALIGNMENT MEASUREMENT IN PROJECT 13

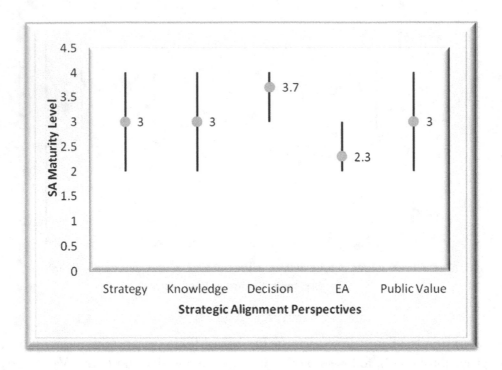

Perspectives maturity: The figure shows that the Decision-Making perspective has the highest average maturity level, Improved (3.7), while the EA perspective has the lowest average maturity level, Committed (2.3). The figure also indicates that the remaining three perspectives, Strategy, Knowledge and Public Value, have an average maturity level of Established (3). Interestingly, of the fourteen projects investigated in this study, only in Project 13 do most of the perspectives receive an attribute maturity level of at least Improved (4), indicating the overall maturity level is high.

As Figure 5-13 indicates, for the second time the Organisation Emphasis on Knowledge attribute received a maturity level of Improved (4) (See also Table 6-1). It should be noted, however, that while the K perspective shows its best maturity level achievement (3) in this project, the EA perspective remains a challenge. When participants were asked how outsourcing decisions were made in regards to EA and its application, they outlined the following challenges contributing to the poor maturity of the EA perspective.

1. Technology-minded decision-making:

[Decisions are] again made by technologists through their influence of the IT Steering Committee. The steering committee made the decisions based on recommendations made by technologists, not from the business. And they certainly don't take a risk management approach to that. (Participant 1)

2. Lack of alignment between technical solutions and business solutions:

We just have to do as Gartner says. We got a very high score on morale; therefore, you should do this thing, but they are not doing risk analysis very well and they don't look at comparative benefits from a business point of view, so you end up with a whole lot of 'technology for technology's sake' type projects rather than projects that drive business transformation and have value add[ing] components. (Participant 1)

The biggest challenge is the fact that the business planning and the business strategic stuff isn't there, it's immature and so that's the biggest challenge... so that's problematic, but having said that, there is a valid role for IT to work with the business, to proactively help the business understand how IT can enhance their business.(Participant 2)

3. Lack of integration between application and technology:

When you have 300 or 400 systems in the backend and no integration, it's very hard to do online successfully. (Participant 9)

The participants' responses above and as the results shown later in Figure 7.4 illustrate that this is perhaps the area where the Council requires the most improvement.

Attributes maturity: In this project, all attributes were spread over three levels of maturity (2, 3 and 4).The project received no Ad Hoc (1) maturity level for any of its attributes. Project 13, like Project 8, has five attributes receiving an Improved (4) maturity level. With the exception of EA, the overall of project maturity is high.

The project has perspective maturity levels ranging from Improved (3.7) to Committed (2.3) and attributes maturity level ranging from Improved (4) to Committed (2).The details of attributes maturity are given in the Table 5-13 in Appendix I.

5.2.14 PROJECT 14

Figure 5-14 shows a summary of the maturity levels of all SA perspectives in Project 14.

FIGURE 5-14: ALIGNMENT MEASUREMENT IN PROJECT 14

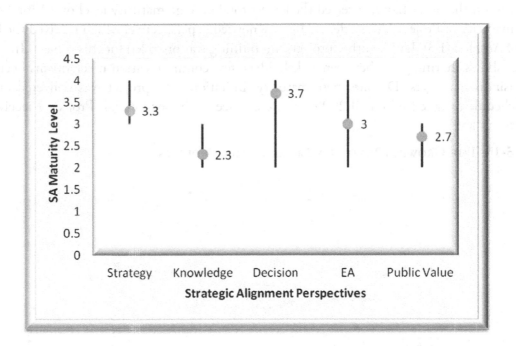

Perspectives maturity: The figure shows that the Decision-Making perspective receives the highest average maturity level, Improved (3.7), while the Knowledge perspective receives the lowest average maturity level, Committed (2.3). The Decision-Making perspective received a high maturity level due to its attributes maturity level, specifically prioritisation and stakeholders, both of which received an improved maturity level (4). The Knowledge perspective, in contrast, must still be addressed. Communication between IT and business and the alignment of technical solutions with business solutions are both already among the priorities of the organisation; both attributes receive a maturity level of Committed (2).

It's really about the awareness. It's more about engaging and it's not just a one-off thing. It's a constant engaging with business and IT around the awareness of the fact that there isn't an IT project, there's a business project that requires an IT component and that the IT bit only enables the business outcome. So it's increasingly just constantly giving that message; that's one really important thing. (Participant 2)

Attributes maturity: In this project, all attributes maturity levels are spread over three levels in the middle (2, 3 and 4).The figure also shows that the project has three perspectives (S, DM and EA) with attribute(s) that received a maturity level of at least Improved (4). In short, the project has perspectives maturity levels ranging from 3.7 to 2.3 and attributes maturity levels ranging from 4 to 2.

In this chapter, fourteen projects were analysed against maturity criteria from Ad Hoc (1) to Optimised (5) maturity levels. All five perspectives and fifteen attributes were addressed in order to discover the SA maturity level of IT projects. The next section presents a summary of findings.

5.3 SA MATURITY SUMMARY: ANALYSIS WITHIN PROJECTS

Figure 5-15 shows the SA maturity level of fourteen projects. The projects are arranged in chronological order, with a total duration of around five years. Project 1 is the earliest project and Project 14 is the most recent. All projects were ongoing between years 2004 and 2009. The horizontal line indicates projects and the vertical line indicates maturity level. The project maturity levels range from 3.1 to 2.1. Project 8 receives the highest total average maturity of 3.1, equivalent to a 61% maturity level. This project stood out because of its direction, performance measurement and quality plan were the best among all projects. Such features are characterised by strategic vision of the project.

Project 6, on the other hand, received the lowest total average maturity level of 2.1, or a 43% maturity level. The maturity of Project 6 was affected by its Knowledge perspective, which received the lowest average maturity of Ad Hoc (1.3). Unlike other projects, no training was provided for this project. In addition, while this project shows sharing goals between stakeholders, few communication mechanisms occurred between IT and business managers. Despite these maturity limitations, the project was delivered on time by the assigned schedule assigned (Council 2009). As we can see in the section 5.4, Project 9 receives the second lowest maturity level.

FIGURE 5-15: THE GROWING MATURITY LEVEL WITHIN PROJECTS

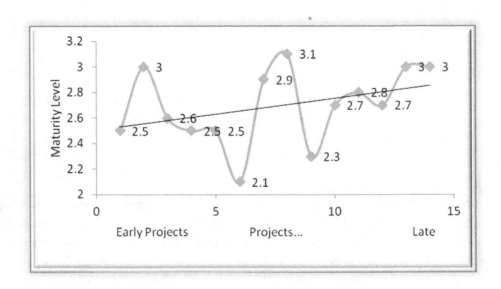

Of the fourteen projects, only two projects (Projects 6 and 9) demonstrate less than 50% maturity level, with an average maturity level of 2.2. The remaining twelve projects demonstrate maturity levels of equal to or more than 50%. The combined projects have an average maturity level of 2.6. Moreover, the findings indicate that the overall SA maturity level increases over time. For example, the average maturity level of the first seven projects is 2.6, compared to the average maturity level of the last seven projects, 2.8. This might be interpreted that the efforts taken by top management and the CIO to increase the overall maturity level of organisation have been successful:

I know, that's what I said you before, we are on a maturity journey, so you start off where you don't have business cases, then you have them, but they are disconnected... that's why X[19]has got such a big job in trying to sort of bring along that discipline in the business more...so that's part of X's role, to try and increase the level. (Participant 2)

Additionally, participants have commented on the role IT governance plays in helping an organisation improve maturity over time:

We have got a thirty-year vision, we have got a five-year corporate plan, we have got a range of corporate strategies which are delivering on those, so we have got quite a lot of mechanisms in place to make sure that the programs and projects we are delivering, are delivering the right things ... so we have now got a lot more strength in our current corporate strategy, so a lot of our outcomes have already been identified. We know what they are for our corporate plan for the next five years in the bold future vision, so when we are developing strategies now, whether it's ICT or other things, they are really not developed in a vacuum; they are being developed and so really the outcomes of those are already known. (Participant 10)

However, it is unknown whether this growing maturity will increase the success rate of projects. To answer this question, further empirical analysis is required and addressed in Section 5.4.

5.4 MEASURING SUCCESS

Based on the definition of success used in this research and the success factors and their attributes values (see Section 4.4), Table 5-1represents the defined success factors and attributes for all fourteen projects. In this table, the four success factors identified are Budget (B), Time (T), Scope (S) and Value Realisation (VR). All factors can be described as being: 'Achieved', 'Mostly achieved', 'Partially achieved' or 'Not achieved'. The table represents the attributes of each factor in each project. As a general observation, it is possible to conclude that nine out of fourteen projects (64% of all projects) achieve most the anticipated benefits (see the VR column in Table 5-1). Similarly, by looking at the Budget column, it is possible to determine, for example, that Projects 3, 5 and 6 were not successful in budgeting, compared to other projects. However, by definition, the significance of budget (weight) and other success factors such as PV can change the initial observed finding. Hence, translating attributes to attributes values (numbers) and weighting each success factor (B, T, S and VR) becomes necessary to gain accurate results.

TABLE 5-1: PRESENTING THE DEFINED FACTORS AND THEIR ATTRIBUTES

Projects	Budget	Time	Scope	VR
Project 1	Partially achieved	Partially achieved	Mostly achieved	Partially achieved
Project 2	Achieved	Partially achieved	Partially achieved	Achieved
Project 3	Not achieved	Not achieved	Mostly achieved	Mostly achieved
Project 4	Partially achieved	Mostly achieved	Mostly achieved	Mostly achieved
Project 5	Not achieved	Partially achieved	Achieved	Mostly achieved

---

19   For confidentiality purposes the name was kept anonymous

Project 6	Not achieved	Mostly achieved	Partially achieved	Partially achieved
Project 7	Partially achieved	Achieved	Partially achieved	Partially achieved
Project 8	Achieved	Achieved	Mostly achieved	Achieved
Project 9	Achieved	Partially achieved	Mostly achieved	Partially achieved
Project 10	Partially achieved	Partially achieved	Not achieved	Not achieved
Project 11	Partially achieved	Partially achieved	Partially achieved	Achieved
Project 12	Mostly achieved	Partially achieved	Mostly achieved	Partially achieved
Project 13	Partially achieved	Achieved	Not achieved	Achieved
Project 14	Partially achieved	Partially achieved	Partially achieved	Achieved

Table 5-2 presents the meaning of the attributes in a number called 'attribute value'. As mentioned earlier in Table 4-3 in Section 4.4.2, the defined attributes value is as follows: 'Achieved' =100%, 'Mostly achieved' = 75%, 'Partially achieved' =50% and 'Not achieved'=25%. The advantage of converting these attributes to numbers (attributes value) is to calculate the success rate value across four factors, which is the next step.

TABLE 5-2: PRESENTING THE DEFINED FACTORS AND THEIR ATTRIBUTE'S VALUE

Projects	Budget	Time	Scope	VR
Project 1	50%	50%	75%	50%
Project 2	100%	50%	50%	100%
Project 3	25%	25%	75%	75%
Project 4	50%	75%	75%	75 %
Project 5	25%	50%	100%	75%
Project 6	25%	75%	50%	50%
Project 7	50%	100%	50%	50%
Project 8	100%	100%	75%	100%
Project 9	100%	50%	75%	50%
Project 10	50%	50%	25%	25%
Project 11	50%	75%	50%	75%
Project 12	75%	75%	75%	50%
Project 13	50%	100%	25%	100%
Project 14	50%	50%	50%	100%

In calculating the success rate of value attributes across four factors, the first need is to revisit the weighting of the four success factors, as presented in Section 4.4.3.

Budget= α alpha = F1= 25%

Time = β beta = F2= 10%

Scope = γ gamma = F3 = 15%

Value Realisation = Ω mean = F4 = 50%

The total weight = 100%

The following formula is used to calculate the success rate of four factors:

$$F = \sum_{i=1}^{4} F_i = 100$$

1.

where F_i refers to shares commonly known as α, β, γ *and* Ω.

Therefore, $\alpha + \beta + \gamma + \Omega = 100$

Thus, taken an example of Project 1:

$$\alpha = 0.5 * 25 = 12.5$$

$$\beta = 0.5 * 10 = 5$$

$$\gamma = 0.75 * 15 = 11.25$$

$$\Omega = 0.75 * 50 = 37.5$$

$$Success\ Rate = 0.5 * 25 + 0.5 * 10 + 0.75 * 15 + 0.5 * 50 = 54$$

54%=0.54

Thus, 0.54 is the success rate of Project 1. The last column in Table 5-3 represents the overall success rate of each project. The range of total success rate varies from 0 to 100.

TABLE 5-3: THE OVERALL SUCCESS RATE OF ALL PROJECTS

Projects	Budget	Time	Scope	PV	Success Rate
Project 1	12.5	5.0	11.3	25.0	0.54
Project 2	25.0	5.0	7.5	50.0	0.88
Project 3	6.3	2.5	11.3	37.5	0.58
Project 4	12.5	7.0	11.3	37.5	0.68
Project 5	6.3	5.0	15.0	37.5	0.64
Project 6	6.3	7.5	7.5	25.0	0.46
Project 7	12.5	10.0	7.5	25.0	0.55
Project 8	25.0	10.0	11.3	50.0	0.96
Project 9	25.0	5.0	11.3	25.0	0.66
Project 10	12.5	5.0	3.8	12.5	0.34
Project 11	12.5	7.5	7.5	37.5	0.65
Project 12	18.8	7.5	11.3	25.0	0.63
Project 13	12.5	10.0	3.8	50.0	0.76
Project 14	12.5	5.0	7.5	50.0	0.75

Table 5-3 indicates that Project 8 received the highest success rate of 0.96, followed by Project 2, which received an 0.88 success rate. In fact, Projects 2, 8, 13, and 14 are the projects that achieved all benefits (on track). Project 10, on the other hand, received the lowest success rate of 0.34, followed by Project 6, which received an 0.46 success rate. Project 10 is characterised by many deficiencies described by PIR internal auditors (Council 2009, p.9) as:

At the time of project inception, the project management plan template was not structured to facilitate ongoing use of the document during the life of the project and did not include all subject areas nor sufficient details in the listed subject areas to adequately describe project governance, standards, performance measurement, management processes and procedures... The project was adversely impacted by the lack of documented project process and governance guidelines... Roles and responsibilities of project members were either not known or not applied satisfactory. A project 'kick-off' workshop was not undertaken. Had it been, this may have alleviated the knowledge gaps of project members.

However, other staff believe there was a really compelling business case that provided clarity, but the main deficiency was due to the misleading actions of the vendor, who demonstrated the project in a simulated environment not in a live environment. When the project was implemented:

It could not do what the vendor said that it could do, so the vendor basically lied... The good thing about that is that governance was so strong on that project, stopping projects is very harder than starting them. (Participant 2)

While providing enough accurate details of a project is vital to its success rate, vendor management becomes an issue and should be included in the IT governance practices.

The study reveals that the overall success rate of projects has improved over time. For example, the average success rate of the first seven projects is lower (0.61) than the average success rate of the last seven projects (0.68). The implication of the study is that the organisation has improved its project management aspects through its SA maturity level in the past few years and this improvement is evident in the project success rate.

This study's research statement (see Section 1.4) states that the purpose of this research is to investigate the impact of Strategic Alignment on government IT projects. Two main aspects of the research have already been addressed: maturity levels, which were addressed in Section 5.2, and success rates, which were addressed in Section 5.4. The central question - how these streams relate to each other—addressed in RQ (a) is now discussed in Section 5.5.

5.5 SA MATURITY AND SUCCESS RATES: ANALYSIS WITHIN PROJECTS

Table 5-4 shows the findings of the success rates and maturity levels of all fourteen projects. The overall maturity pattern is shown to relate strongly to the overall success rate pattern. An example is Project 1: when its maturity level increases from 2.5 to 3 and then decreases to 2.6, its success rate also increases from 0.54 to 0.88 and then decreases to 0.58. Similarly, when Project 7 increases its maturity from 2.9 to 3.1, its success rate also increases from 0.55 to 0.96. Likewise, when Project 10's maturity level increases from 2.7 to 2.8, its success rate also increases from 0.34 to 0.65. Similar patterns also occur in Project 12. The implication of this finding is that the SA maturity level has a strong impact on the project success rate. These findings correspond to the previous literature where alignment improves performance (Tallon and Kraemer 2003; Luftman 2011) through decision-making on IT-related investments (ITGI 2008).

TABLE 5-4: RELATIONSHIP BETWEEN SA MATURITY LEVEL AND SUCCESS RATE

Projects	Success Rate	Maturity Level
Project 1	0.54	2.5
Project 2	0.88	3
Project 3	0.58	2.6
Project 4	0.69	2.5
Project 5	0.64	2.5
Project 6	0.46	2.1
Project 7	0.55	2.9

Project 8	0.96	3.1
Project 9	0.66	2.3
Project 10	0.34	2.7
Project 11	0.65	2.8
Project 12	0.63	2.7
Project 13	0.76	3
Project 14	0.75	3

Figure 5-16 portrays the relationship between success rate and maturity level in a low to high success outcome order. The model linear data was used to interpret data points and trends as most project management data sets. The notion of line of best fit model allows researcher to construct scatter plots of two-variable data such as success and maturity, interpret individual data points of each project and make conclusions about trends in data, especially linear relationships and estimate equations of lines of best fit.

The findings correspond to the literature research where strategic alignment plays a vital role in controlling and overseeing the IT activities that lead to project success (Nolan & McFarlan, 2005; Cresswell et al 2006). As the graphic indicates, Project 10 receives the lowest success rate of 0.34 and Project 4 receives the highest success rate of 0.96. As the level of maturity increases, the success rate increases. This finding shows the impact of SA on projects' success rates. Therefore, the case study demonstrates the linkage between SA and project outcome and shows how this outcome can be affected if SA has not been taken into consideration.

FIGURE 5-16: RELATIONSHIP BETWEEN SUCCESS AND MATURITY OUTCOME

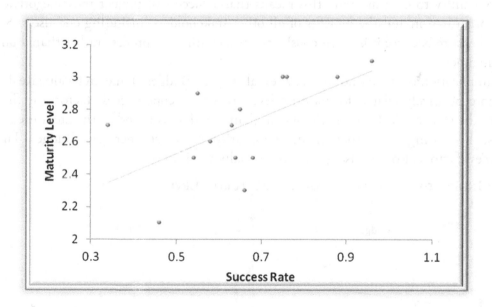

Finally, there are a few projects that show a contrast between maturity level and success rate, such as in Projects 3 and 9. The findings from these two projects compared to the overall success rate of all projects may reveal that the IT project team in Project 3 work to their best when its maturity decreases, however the IT project teams did not consider increased SA maturity in Project 9 to their own advantage, resulting a lack of connectivity between IT project staff and top management who work for the business's sake.

Next, the relationship between individual success factors (known as Budget, Time, Scope and Value Realisation) and maturity level is presented.

5.6 RELATIONSHIP BETWEEN INDIVIDUAL SUCCESS FACTORS AND SA MATURITY LEVEL

Section 5.5 presented information on how four success factors (B, T, S and VR) together relate to the maturity level of fourteen projects. Figure 5-17 shows how these individual factors relate to the maturity level. As Figure 5-17 indicates, the Budget, Time and Value Realisation success factors seem to have a strong relationship to the maturity level, whereas Scope success factor has a negative relationship to the maturity level. It is very interesting to investigate why a Scope success factor has a contradictive result with maturity level. A first possible explanation could be found in the fact that project must be completed 'with minimum or mutually agreed upon scope changes' (Maylor, 2005, p.288) which was not the case in projects investigated, even though variation is accepted by stakeholders due to the nature of project changing environment. Consider the following participant (Participant 4) who argues to minimising variations in project:

> 'Some strategic projects run over 3 years and if you were trying to minimize change in the project for the sake of it you might miss a very valuable opportunity'.

A second possible answer could be the fact that traditional scope framework implementation of checking off tasks and milestones on the project schedule was not enough to meet addition of qualitative objectives rather than quantitative. 'the scope has to deliver to the quality required by the client to be considered a success' (Participant 7). A confirm scope methodology is required that different group of people will have benefits differently from the project which includes efficiency during the project, effectiveness to customers and future opportunity to organisation. This means that a successful project must negotiate between the benefits of the organisation and the benefits of all other stakeholders, including end-users. Such success is always hard to measure because it is often considered as a continuous process rather than a pin point at the end (project outcome).

While the implication of these findings may reveal that the Budget, Time and Value Realisation success factors were more positively related to maturity level than the Scope success factor, the findings are not contradictory to the fact that VR is more significant than B and T, as found in the literature and interviews. In other words, the findings show that when the maturity of projects increases, Budget, Time and Value Realisation success factors also increase, yet are not equally significant.

FIGURE 5-17: INDIVIDUAL SUCCESS FACTOR AND MATURITY LEVEL

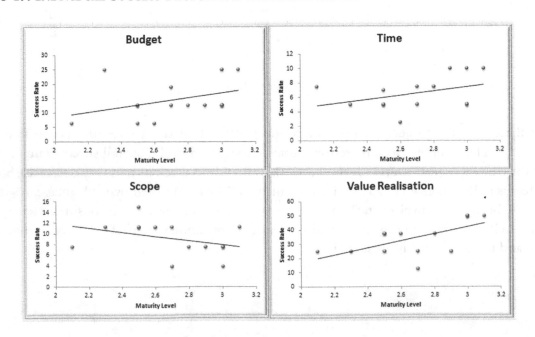

5.7 Conclusion

This chapter presented the findings of the case study conducted in a local government. By addressing the research questions RQ(a), *What is the relationship between the SA maturity level and the success rate of IT projects?*, the chapter provided an analysis of the case study to determine whether SA maturity affected government IT projects' success rates. The chapter therefore has outlined and discussed the contributions to the knowledge and literature for the study.

By examining the maturity level, the findings indicate that the overall maturity level of fourteen projects has improved. The overall maturity level of projects ranged between 3.1 and 2.1. The study also shows the success rate of IT projects improved over time. The findings in RQ(a) indicate that maturity level of SA positively facilitates the projects' success rate. The findings confirm the general prediction in the literature that SA perspectives are important to the project success, thus supporting the conceptual model. Until now, the study has not shown which perspectives or attributes contribute more to the project success. In other words, there is a need to find out which perspectives or attributes are more significant than others to the project success, leading us to further query RQ(b) and RQ(c) in Chapters 6 and 7 respectively, which present the across-projects analysis results.

6. Analysis of SA Attributes and Maturity Levels Across Projects

6.1 General Remarks

Like Chapter 5, Chapter 6 deals with the impact of SA maturity levels, but in particular it is concerned with the effects of attributes on the success rate of IT projects. This chapter consists of six sections. Section 6.2 provides the overall findings of attributes maturity level. Sections 6.3representsthe across-projects analysis of the relationship between attribute maturity and project success rate. This section answers the RQ (b) question: *'Which SA attributes promote success of IT projects?'*, followed by a summary section, Section 6.4. Section 6.5concludes this chapter.

6.2 attributes Maturity Levels: Across-Projects Analysis

Table 6-1 presents the total maturity summary of SA perspectives and attributes of all fourteen projects. In this chapter, only SA attributes maturity levels are presented and analysed across projects. Overall, the Clarity of Direction attribute receives the highest average maturity level of 3.4 across all projects, followed by the IT Investments and Budget and Benefits to Organisation attributes which both receive a 3.1 average maturity level. In contrast, the Economic/Financial Metrics attribute receives the lowest average maturity level of 1.9, followed by an average of 2.0 for the Training attribute.

As Table 6-1 indicates, the attribute maturity levels range between 1 and 4. Three projects (Project 2, 8, and 13) have an Improved maturity level (4) in five of their attributes. Similarly, three projects (Projects 1, 6 and 9)have an Ad Hoc maturity level (1) in two of their attributes. Among all attributes, Clarity of Direction received the highest maturity level of Improved (4) in five projects, followed by Benefits to Organisation which received the same maturity in four projects. On the other hand, Training received the lowest maturity (1) in five projects followed by Economic/Financial Metrics, which received a 2 maturity level in all projects except two (Projects 3 and 9) which have the lower Ad Hoc maturity level (1). Details on these attribute maturity levels and their relations to success rate are provided below.

TABLE 6-1: MATURITY LEVELS IN THE FOURTEEN PROJECTS

Perspectives	Attributes	Proj1	Proj2	Proj 3	Proj 4	Proj 5	Proj 6	Proj 7	Proj 8	Proj 9	Proj 10	Proj 11	Proj 12	Proj 13	Proj 14	Average across projects
Strategy	Clarity of Direction	4	4	3	3	4	3	4	4	3	4	3	3	3	3	3.4
	Performance Measures	1	3	3	2	2	2	4	4	3	3	3	3	2	3	2.7
	Quality of IT/business plan	2	3	2	2	2	2	3	4	2	2	3	2	4	4	2.6
Average		2.3	3.3	2.7	2.3	2.7	2.3	3.7	4	2.7	3	3	2.7	3	3.3	2.9
Knowledge	IT/business Managers Participation	2	2	2	2	2	1	2	2	2	3	2	2	3	2	2.1
	Organisation Emphasis on Knowledge	2	4	3	3	2	2	3	3	3	2	3	3	4	3	2.9
	Training	1	1	3	2	1	1	3	2	1	3	3	3	2	2	2
Average		1.7	2.3	2.7	2.3	1.7	1.3	2.7	2.3	2	2.7	2.7	2.7	3	2.3	2.3
Decision-Making	IT Investments and Budget	4	4	3	3	3	3	3	4	3	3	2	3	3	3	3.1
	Prioritisation	3	4	2	2	2	3	2	3	3	3	3	3	4	4	2.9
	Stakeholders	2	3	3	3	3	2	3	3	2	2	3	3	4	4	2.9
Average		3	3.7	2.7	2.7	2.7	2.7	2.7	3.3	2.7	2.5	2.7	3	3.7	3.7	3
Enterprise Architecture	Aligned Technical/Business Solutions	2	3	3	2	3	2	3	3	3	3	3	2	2	2	2.6
	Application and Technology	4	3	2	3	3	2	2	3	3	3	3	3	3	3	2.8
	Risk Assessment	3	2	3	2	3	2	3	3	2	3	3	2	2	4	2.7
Average		3	2.7	2.7	2.3	3	2	2.7	3	2.3	3	3	2.3	2.3	3	2.7
Public Value	Benefits to Organisation	3	4	2	4	2	3	3	4	2	3	3	3	4	3	3.1
	Benefits to Public	2	3	4	3	3	2	3	2	3	2	3	3	3	3	2.8
	Economic/Financial Metrics	2	2	1	2	2	2	2	2	1	2	2	2	2	2	1.9
Average		2.3	3	2.3	3	2.3	2.3	2.7	2.7	2	2.3	2.7	2.7	3	2.7	2.6
All attribute average		2.5	3	2.6	2.5	2.5	2.1	2.9	3.1	2.3	2.7	2.8	2.7	3	3	2.7
Total SA maturity sum score		37	45	39	38	37	32	43	46	35	41	42	40	45	45	40
Percentage		49%	60%	52%	51%	49%	43%	57%	61%	47%	55%	56%	53%	60%	60%	54%

116

6.3 Attribute Maturity Levels and Success Rates: Across-Projects Analysis

Clarity of Direction: The Clarity of Direction attribute received the highest maturity level of Improved (4) and the lowest maturity level of Established (3). In this attribute, 43% of the projects have an Improved maturity level. None of the other attributes have a percentage this high in this maturity level. The remaining 57% of the projects have an Established (3) maturity level (See Appendix J[20]). This mean the attribute has neither a Committed (2) nor Ad Hoc (1) maturity levels in any of the fourteen projects. In this attribute, the average maturity level across projects is 3.4. The findings indicate that Clarity of Direction received the highest average maturity level among all attributes, across projects, followed by the IT Investments and Budget and Benefits to Organisation attributes, which both received a 3.1 maturity level.

The high maturity of Clarity of Direction among all attributes is due to the government commitment towards its direction.

I guess in terms of delivering on the vision whether it's the ICT vision or the organisation's vision, it's about that making sure people understand.

(Participant 10)

Clarity of Direction is one of important elements of the Strategy perspective and organisation has paid clear attention to it.

You don't do any implementation work until you have clear, understood business plan, everyone knows that's what you are doing. (Participant 3)

The two questions that arose the most when vision was to be communicated were the reason that the project needed to be carried out and how the new changes would benefit the organisation and society. The overall objective of the Strategy is to increase the clarity of direction and purpose of information and technology investments(Council 2006).

Such findings are supported by other studies, which place emphasis on the clarity of direction from high-level executives and managers responsible for the ICT strategy (Fardal 2007; Dodd, Yu et al. 2009).Most of the objectives and the direction of the projects studied here were clear and well communicated, which made them easier to be accepted and implemented by systems owners. Without presenting a clear objective from very beginning of the project's implementation to stakeholders, the project will suffer later on.

Despite these findings, which show that there was a plan to make sure that all of the project's objectives are clear, the attribute contribution to the project success rate was low (See Appendix J[21]). This problem could be due to the fact that while the Executive Leadership Team (ELT) were clear in their direction, there was a gap in cooperation between other project team members, particularly between business and IT members. This problem needs to be addressed, as expressed by one of participants:

We are trying to develop [a] service profile and then hopefully mature that business planning as well. We got really mature ICT planning and immature business planning. You get dysfunction. We need to get IT and business together.(Participant 5)

Performance Measures: The Performance Measures attribute received the highest maturity level of Improved (4) in two of the fourteen projects and the lowest maturity level of Ad Hoc (1) in one of the projects. The Ad Hoc maturity level occurred in Project 1. The projects began over a five-year period, and as Project 1 occurred in the earliest year among the projects reviewed and the results in the subsequent projects have a higher maturity level for this, the findings indicate that the Performance Measures attribute became more

---

20   The graphic percentage of maturity level distribution of each attribute across projects is presented in Appendix K

21   The relation of each attribute to success rate is described as positive, moderate or negative. All graphics depict this relation from here on ward are presented in the Appendix J.

mature in subsequent projects. This improvement of maturity in this attribute helped to minimise risk and improve benefits in later projects. The maturity level of this attribute is Established (3) and was found in 50% of the projects. The Committed maturity level (2) was found in 29% of projects. The Performance Measures attribute has a total of four maturity levels as indicated in Figure 6-2 in Appendix J.

The findings from the case study reveal that this attribute focused on information that controls and secures the projects. The types of information required included risks, cost estimation, milestones, deliverables, acceptable levels of alerts, key performance indicators, reports, baseline plans, variation, measurement tools, in/out scope, constraints/dependencies, project structure and value realisation plan. The availability of this information in BC and PMP documents varied from one project (for example, Project 1) to others (for example, Projects 7 and 8) depending on the availability of information needed during the creation of BC and PMP documents and on project types. Project 10, which was a large project, required more information and baseline data than a smaller project (such as Project 9). The study reveals that such information forms the basis for guaranteeing the success of a project. If such information is missed in the project, success only occurs by chance.

The findings, however also reveal that the organisation still faces challenges in evaluating its IT project performance. When the question 'What do you think are the main organisational challenges or weaknesses in evaluating the performance of IT?' was asked, the answer was generally positive and the response was similar among the participants:

I don't think they do. So I think they are only just starting there to look at how ICT performs within their business and when we do in new projects. So I think they only just started. I think they only started benefits realisation.

(Participant 7)

..again we take technological view, a very narrow technological view. rather than a business view so that's our primary weakness. The other weakness we have...We don't do benefits realisation and the other issue is we allow people who are doing projects to evaluate themselves so when the Project Manager who's being paid get to mark themselves whether or not they did a good job. They generally said they did a good job, which isn't very smart.

(Participant 1)

Quality of IT/Business Plan: The Quality of IT/Business Plan attribute received the highest maturity level of Improved (4) in 21% of the fourteen projects (Projects 8, 13 and 14). These three projects were situated in the second half of the projects (implemented in the final years surveyed), indicating that project maturity levels are better in later projects than in earlier ones. This attribute received the lowest maturity level of Committed (2) in 57% of the projects while the remaining projects (21%)have an Established maturity level (3). The total average maturity level of this attribute across all projects is 2.6.

The Quality of IT/Business Plan outlines business requirement details, including the need for change, desired outcomes, scope analysis, value profile, risks/issues, budget/cost estimate and assumptions/constraints, which were addressed during project implementation. The findings indicate that this attribute is a crucial part of project process and, if not well defined, can create problems during and after implementation, as the attribute can affect the anticipated benefits. In this study, the findings reveal that this attribute promotes project success (Figure 6.1). This finding is supported by other literature. For example, Matt Howell, Head of Public Services, Technology at Cap Gemini Institute, United Kingdom, thinks that many government IT projects are doomed to failure before they even begin and argues that the problem will not be solved without an improved approach to procurement (Jones 2009).

FIGURE 6--6-1: RELATIONSHIP BETWEEN SUCCESS RATE AND MATURITY LEVEL

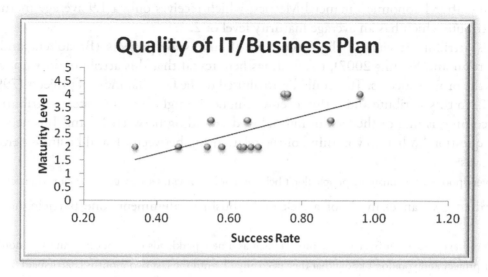

Thus the absence of a quality IT/business plan is often a key factor in project failure. In addition, the quality of an IT/business plan cannot be judged by looking at BC without also looking at other documents such as PMP or the Project Charter. While project failure can be the result of many factors, the findings here show that a missing high-quality IT/business plan (as in Project 10) increases the chances of project failure. This could be the result of missing the accurate and detailed information required.

At the time of project inception, the project management plan template was not structured to facilitate ongoing use of the document during the life of the project and did not include all subject areas nor sufficient details in the listed subject areas to adequately describe project governance, standards, performance measurement, management processes and procedures.

There were two periods in the history of Project 10 where the project plan was significantly impacted by unplanned situations.

The first of these occurred after the tender was awarded and it was discovered the prime vendor did not have adequate documentation to enable a satisfactory analysis of the fit between the functionality offered and the business requirements. This led to the execution of a Heads of Agreement between the parties to develop the missing documentation.

The second period of unplanned activity occurred mid to late 2006 when doubts regarding the capability of the prime vendor arose, and the timeframe and costs of delivering the preferred solution were projected to exceed reasonable tolerance levels'.(PIR internal auditors, CGC 5th August Council (2009, p.7)

In addition, another major challenge that the Council faces in producing good quality IT/business plans isthat good business requirements are hard to document. Such documentation is both time and money-consuming and business participation isalso required (it is not merely an IT focus).

..my point of view, the minus project is only a distance I've even worked at it. It didn't start as with well-defined business requirements - that was its killer problem. Because it did not have defined business requirements I think they almost come up to the market and say, "What have we got up there?", and what's is working all become very difficult. If they've had good business requirements the project would have been successful. (Participant 7)

So these business cases are taking a year, 18 months to put together and they are still not well formed and they end up being huge. I have seen a 140-page business case… Very hard work.(Participant 9)

IT/Business Managers' Participation: The IT/Business Managers' Participation attribute received the highest maturity level of Established (3) in 14% of the fourteen projects and the lowest maturity level of Ad Hoc (1) in 7% of those projects. The remaining 79% of the projects have a Committed maturity level (2), which is not promising. The average maturity level across all projects in this attribute is 2.1, and is the third lowest maturity attribute after Economic/Financial Metrics, which receives only a 1.9 average maturity level and the Training attribute which has an average maturity level of 2.

Though this attribute is vital to the success of all large-scale projects (Broadbent and Weill 1997; Evans 2004;Gartlan and Shanks 2007), the findings here reveal that this attribute does not show a strong contribution to the project success. This could be attributed to the fact that most of projects (79%) have a low maturity level (2) in this attribute and if this level was higher it might have a greater contribution to success. Its major shortcoming is due to the lack of mutual understanding between IT and Business managers. In response to the question 'What do you think of the relationship between IT and business here?' the answer, unsurprisingly, was:

> Pretty poor, very poor… The business people don't believe that IT provides a very good service. (Participant 9)

When asked to give an example of a lack of IT/business alignment, one participant responded by saying:

> We are going to have a major infrastructure project such as a new parklands or something and you don't know, that's a two- or three-year project and you don't know that they need any IT until the last two months.(Participant 3)

The findings support the findings in other studies. Evans (2004) reports that business experts have a negative perception of the IT function. The business sector regards IT professionals as 'strange' and 'part of the problem'. Fielding (2002) also notes that most IT departments are still held in low esteem by their business counterparts.

Participants believe that the participation of IT/business managers can be improved by better defining the expected business change. This can be achieved by categorising the business into 'as is' for the current situation and then 'to be' for how it should be in the future. This can be achieved by helping the business to identify their issues and solutions at the very beginning of the project and talking to ICT people in ordinary language (non-technical language) about what business changes and benefits can be brought by ICT.

Organisation Emphasis on Knowledge: This knowledge attribute receives the highest maturity level of Improved (4) in 14% of the fourteen projects and the lowest maturity level of Committed (2) in 29% of the projects. The remaining 57% of the projects have an Established (3) maturity level. The average maturity level in this attribute across all projects is 2.9. This average maturity level was accurately predicted by Participant 2, when he/she said "I would plot at 2.5 to 3".

The findings prove this attribute to be significant to the entire organisation. Clear communication is a necessity for project success. All available channels of communications were used in the successful projects, including internal letters, emails, websites, project meeting minutes, a detailed communication plan, corporate reporting across the organisation and a discussion forum. That climate of knowledge-sharing increased the understanding between stakeholders and facilitated project implementation across various departments in the organisation. The findings indicate that this attribute has a positive impact on project success which is in agreement with the research literature, which reported that strong recommendations are needed for cross-boundary collaborations and government information-sharing (Green and Ali 2007; NASCIO 2007; NASCIO 2008; Soeparman, Duivenboden et al. 2009).

> Quality of performance and speed are also increasing due to increased communication within and among project teams. Results are being achieved by applying timely, effective, and interactive, communication.

> (Carr, Folliard et al. 1999, p.172)

The credit for this impact is also given to the reporting mechanisms deployed in the organisation that ensure the effectiveness of the life cycle of the IT project.

> We do lots of reporting now. We have done some really good work in the last couple of years; you know we are using our project management software. We are actually reporting that project status report now. Every month we actually have a standardising on deliverables so we can actually do composite reporting across deliverables. Before that every project manager

have called anything what they liked, did whatever they have liked, now we are putting standards in so we can do combined reporting and we have proper phase gates and we have phase gate really. So we are actually getting quite good at that now. It was effectiveness; in fact this is the life cycle. (Participant 9)

While the processes of these IT projects were well reported via various standard mechanisms, there was some miscommunication between the organisation and the stakeholders in some areas at the beginning of the process, specifically in the concept stage when the project was first deemed to be considered as the project. In other words, no one from the organisation contacted the customers to tell them that their concept plan had not been accepted as a project, hence it was re-classified from a project to a major task.

So the role is about who contacts who once a concept (project) is seized is fairly unclear. And I think the organisation as a whole and business stakeholders are not clear what it is going to happen once they put the concept plan in.

(Participant 8)

Training: At different levels, the importance of the IT program and project training among staff and stakeholders has been supported by previous IS research (Fielding 2002; EDP 2005; Khosrowpour 2006; Baxi and PSVillage 2007; EDP 2007; Anisetty and Young 2011). In this case study, that training was supported by the top management of the organisation. For example, training of users and support personnel was mentioned in the organisation's strategy goal. Some details of that training were given in Project 3; for example, when training would commence and finish and the users' acceptance test. In addition, training was targeted at certain groups, such as the decision-making members or the relevant staff who were required to activate Project 4 and Project 7, the across-organisation selection of 120 staff for test deployment of the Project 11 and the train-the-trainer and kick-off workshops in Projects 12 and 14.In fact, the organisation, though immature, has already taken step towards implementing training.

A part of that was to establish almost—we call it the centre of excellence, but a unit which is responsible for training, capability, quality and all those things.

(Participant 10)

However, the study shows that training was poorly managed and monitored. As a result, workshops were not attended, especially with those projects run in the early half of the study (the first seven projects).The study also shows that this attribute received the second lowest average maturity level (2) across the fourteen projects, followed by the Economic/Financial Metrics attribute (1.9). Training is the only attribute that had an Ad Hoc maturity level (1), in five of the projects. Unfortunately, the Training attribute was not shown to be significant to the success of the projects studied. The findings in this attribute contrast those for the Organisation Emphasis on Knowledge attribute, where if a lot of reporting mechanisms are used, it enhances the understanding of the project implementation processes amongst diversity of stakeholders.

The issues surround the organisation's training need to be addressed in many organisational areas as they are needed to improve the current maturity level. This finding was supported by more than one participant.

..follow the corporate risk management standard, get trained how to do enhance risk management, how to understand risk and mitigate them and do that up front before you buy the product. (Participant 1)

Training, top-down leadership, communicating, engaging, allocating resources.

(Participant 8)

IT Investments and Budget: This attribute is a core element of many projects and is very important as without the right decision, funding may be allocated to the wrong place right from the beginning, resulting in a project that delivers no value to the organisation. In this organisation, approval of budgets on IT investments is normally undertaken in a formal process through corporate governance committee. Such a process is followed by many organisations (Weill and Ross 2004).

The process of making formal decisions is resource and time-consuming in an organisation which requires details of project proposals which sometimes need to be submitted in two different methods, on both project charter and business case forms. Most formal decisions were made based on two streams: budget and IT value. Thus, information such as cost analysis, directorate/project personal costs, risk contingency, ongoing support cost, cumulative cost/benefits, alignment position/priority to ICT strategy and corporate plan, value profile and inputs from stakeholders and value realisation plans are amongst the key elements for consideration with any project approval. Such efforts contribute to its maturity. The findings show that this attribute has a strong impact on project success rate. The total average maturity of IT Investments and Budget attribute's across projects was the second highest (3.1) after Clarity of Direction (3.4). In addition, 93% of projects have an Established (3) or Improved (4) maturity level in these attributes (Figure 6-2).

FIGURE 6-2: IT INVESTMENTS AND BUDGET: ATTRIBUTE VALUE ACROSS PROJECTS

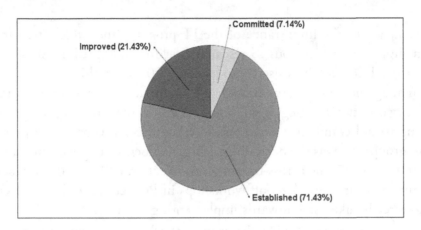

However, in the assessed organisation the attribute was not free from challenges. One of the biggest challenges is obtaining approval of a project or budget. Participant 3 said that "the other challenge is always getting budget". Participant 8 also commented that:

There is a definite linkage missing between the governance and the financial management, so we can get initiatives that are approved to go ahead and have no dollars or we can end up having dollars with no approval to go ahead, so there is a mismatch there and we all know it...(Participant 8)

Campbell (2003) also noted that government agencies don't have the financial flexibility to quickly cope with rapid changes of technology like private sector organisations do. This is mainly due to budgets being planned at least a year in advance while systems may have to respond in the short term as a result of a ministerial request. Having to deal with legislative requirements and the expectations of citizens in a changing environment adds to the complexity of system delivery and hence to the achievement of IT/IS project success.

This challenge, however, can be understood where managing cost/benefits profiles, risk analysis and priority position to organisation is needed to avoid undesired outcomes (NASCIO 2007; Soeparman, Duivenboden et al. 2009). This is because the impact of decisions can be observed at more than one level: the organisation, business, stakeholder and IT project levels and in the IT project outcome (Aurum, Wohlin et al. 2006). Generating IT value requires a strong foundation of quality information as well as a clear understanding of that information among team leaders and program and project management members. Project 10 in the case study lacked such vital information and this resulted in negative consequences.

One effective way to improve budget decisions of IT/IS processes is to justify their benefits (IT value), as expressed by one of the participants, and a second is to prioritise projects according to the corporation's ICT strategy and priority.

...the question is how is that going to affect members of our community and if nothing is going to happen, if it doesn't make any difference then why would you want to do it.(Participant 10)

Prioritisation: The Prioritisation attribute received the highest maturity level of Improved (4) in 21% of the fourteen projects and the lowest maturity level of Committed (2) in 29% of the projects. The remaining 50% of projects have an Established maturity level (3). The average maturity level in this attribute across all fourteen projects is 2.9.

Since there were insufficient resources to undertake everything simultaneously, prioritisation was vital. In this case study, prioritisation reduced costs, eliminated the need for endless meetings and created the maximum value for the resources available. Prioritisation across projects has been shown to contribute positively to project success as priority and value for money is considered as an IT governance principle and is important to an organisation. Thus, the Prioritisation attribute helps an organisation to make the right decisions on project investments for IT value and to choose projects that are more aligned with their corporate strategy. In addition, using tools to measure IT value enhances the maturity of this attribute. For example, when the question 'How does the tool such as BVIT to measure IT/Business value work?' was asked, the answer was:

It works well.…. Probably four (at 4 level)…. People may want to improve it, but it's allowing us to prioritise our portfolio and in the three or so years that we have had it implemented no one has actually asked to change it.(Participant 3)

On the other hand, as acknowledged by Fardal (2007, p.1), "Well-formulated ICT strategies make it possible to identify and define the right ICT projects, however defining the right projects do not imply that the projects will be successful". The challenges the organisation may face in relation to this attribute was described by one of the participants.

At a lower level, symptoms of misalignment I think when you get too many interim solutions. People sort of say "Yeah, yeah, we know all that,. We know what we need over here but we just need to do a little thing over here, it's not going to cost very much money." They said that. For that little amount of money, it could have done to something else. So that's a misalignment of priorities when the organisation are saying these are our priorities and business leaders are saying, "That's all good but I need my little patch here". What you find there you're getting those resource priorities means that those resources are being spent on things that aren't really strategically aligned. Interims solutions are binding our lives and again that's getting better as well. More and more people are willing to take on a corporate view. And your budgeting and your process emm... and your financial system needs that support as well.

(Participant 5)

In addition, there were other situations where the quantification of the project and project-deferral risks and their impact on value were not easy to include in the process of effective decision-making. For example, senior management faced a dilemma when choosing between keeping Project 10 or winding it down due to the project's deficiencies, resulting in a delayed decision. Such a delay cost the organisation hundreds of dollars daily.

Overall, the results indicate that prioritisation is not only significant to budgets and project selection, but also to project success rate.

Stakeholders: The Stakeholders attribute received the highest maturity level of Improved (4) in 14% of the fourteen projects and the lowest maturity level of Committed (2) in 29% of the projects. The remaining 57% of projects have an Established maturity level (3). The Stakeholders attribute also has an average maturity level across all projects of 2.9, which is the same result as with the Prioritisation attribute.

Stakeholders are important to an organisation. In this study, stakeholders were found to be engaged through portfolio management, corporate activity framework, business engagement and through the life cycles of the projects. However, the study shows that stakeholders have been a source of advantages and disadvantages to the organisation. An example of a stakeholder-based advantage is:

Stakeholders help to reduce a high level of process variation (the life cycle of the projects) and thus, enhance the success rate.

..but the project cannot be successful without the input of these people... because they are the intelligence, they are... [the] heart of the business...You need to get a balance of continuing to run the business whilst putting sufficient intelligence to make it work. That is a very hard challenge. (Participant 4)

But we try to get input from stakeholders through the portfolio management committee. This committee come along and are explained the concept plan to mainly the IT people. And that works well. (Participant 8)

...early engagement and within our idea concept basis to ensure we are going to get a value from them and then carry through in the life cycle, so really managing our risk around benefit delivery and the outcomes that the business require. (Participant 6)

But one issue here, as pointed out by one participant, is:

Are we engaging in the right stakeholders, at the right time with the right information? (Participant 6)

Three examples of disadvantages to the organisation are:

Stakeholders, in particular contractors, know their business, such as project management, methodology and the life cycle of projects, but they do not know the organisation's business, reducing the level of cooperation between them.

We do massive amounts of infrastructure and systems and yet we do not have project management as a core area of the business so what we do, we buy it in. That means we buy in contractors. What don't contractors know? They don't know our business...and that is one of the biggest risks to projects.

(Participant 4)

This relationship can worsen in some situations, as explained by one of participants:

Erm, the relationship between the vendor and council business is not as good as it should be so council withhold information that vendor needs or the vendor misrepresent their capacity because they don't want to lose the contract.

(Participant 1)

Such a problem can extend far enough to encourage vendors to minimise technology costs rather than to maximise business benefits, therefore the chance of a quality relationship is considered low. A greater mutual benefits trust and more cooperation are needed.

Stakeholders are not all engaged over the project's life cycle.

From a macro point of view, as to engaging the end-user or the supplier. They are not factored into those structures. They should be but they are not.

(Participant 1)

However, engaging a lot of stakeholders has not provided a good business result in the past. For example, getting the right stakeholder group has not been an easy task and, thus, did not help in the organisation's business. However, neither does engaging a lot of stakeholders help an organisation to improve its business, as some stakeholders are trying to sell as much of their products as possible to an organisation.

..previously we had so many projects and steering committees, but it wasn't possible to get the right people, the right groups. Attendance was poor, so that was very much left to, I guess the ICT elements in the organisation to make these projects happen which didn't help to engage in the business because we just had too many projects, too many people trying to do too many things.

(Participant 10)

Stakeholders are sometimes negatively affected by immature IT governance management behaviour.

..project managers are not actually necessarily working for the client who is in a different directorate. They work for themselves. And, actually, that's the reason I left them. I found that it is very difficult relationship too... Because sometimes they will be doing things that work particularly good for the client then they want to reduce those change things they do things which the client is fully aware will not be happy with. So they can be quite difficult where Project Managers are unable to really champion the projects to deliver for the clients so I think that's quite a problem.(Participant 7)

Even with all of the challenges discussed above, stakeholders remain a core group and show a positive relation to the success rate of projects. The credit for that status goes to the vital and prominent role these groups play in the different life cycle stages of projects.

Aligned Technical/Business Solutions: The Aligned Technical/Business Solutions attribute received the highest maturity level of Established (3) in 57% of the fourteen projects and the lowest maturity level of Committed (2) in 43% of the projects. The average maturity level in this attribute across all projects is 2.6. This attribute has neither an Improved maturity level (4) nor an Ad Hoc maturity level (1) in any of the projects.

This attribute is useful as a means for collaboration among the business vision, business requirements and information technology parts of an organisation (Kaisler, Armour et al. 2005; Ylimaki and Halttunen 2005/06, p.189). In the case study, an example of this collaboration was seen when the Information Management (IM) program operationalised the 2005–2009 ICT Strategy and aligned the 2005–2009 Corporate Plan and integrated the solution/risk with other projects as well as with directorate impacts by internal resources and information security, as it forms part of the overall Corporate Risk Management framework implementation. Such collaboration eases project implementation and it has been argued that it improves organisational performance (Maes, Rijsenbrij et al. 2000; Zarvic and Wieringa 2004; Hugoson, Magoulas et al. 2010;Simons, A. et al. 2010).The literature reveals the relative importance of this attribute to promoting the success rate of projects.

However, in the case study, the Aligned Technical/Business Solutions attribute has not shown a strong contribution to success rate. The challenges of this attribute are similar to those of the IT/Business Manager Participation attribute. One difference is that the former theoretically derives from the Knowledge perspective and relates to people, whereas the latter theoretically derives from the Enterprise Architecture perspective and relates to technology. The issue of aligning technical and business solutions relies on two aspects: how business can better understand its own business processes and how IT people communicate with business people, not using technical language but instead in business process terms. Thus, this attribute involves people interaction, and technology cannot stand on its own or be studied in isolation. Understanding these two aspects can improve the application of technology.

So in lot of cases the business hasn't got well-documented processes, isn't clear on how it runs its business... So that when there is discussion about improvement, I am not sure what they are improving from and to. So I think that's one thing the business needs to get a better handle on... It's just looking within its own little patch and not looking outside of that. So it's got a lot of blinkered approach, it's looking to fit its immediate problem. (Participant 10)

The solution is simple:

..we need to do upfront when we do the planning for the project, focus on the business, almost not worry about the technological solution and then say, okay what technology is out there that can give us the highest level of fit with our business requirements.(Participant 10)

Application and Technology: The Application and Technology attribute received the highest maturity level of Improved (4) in 7% of the fourteen projects (one project) and the lowest maturity level of Committed (2) in 29% of the projects. The remaining 64% of the projects have an Established maturity level (3). The average maturity level in this attribute across all projects is 2.8.

The Application and Technology attribute in this study was found to include how technological capability and integration match with the business application. Hence, the importance of the Application and Technology attribute's role rose. To address this issue, EA came to the forefront. So far, the organisation has aligned its business plans well to its strategic priorities and corporate plan. "Much better, that stuff is aligning it up, much better" (Participant 9).The study shows the overall maturity level across projects of this attribute is high (2.8) and helps to promote the success rate of IT projects. The findings support findings in the literature.

..unless business and Information Systems management can agree to partner and work together, the organisation is unlikely to achieve its goals.

(Murray and Trefts 2000, p.3)

On the other hand, literature shows that aligning and integrating technology capabilities with business applications (the outcome) have not been easy tasks in the EA of organisations (Kaisler, Armour et al. 2005). This is due to the heavy loads on the systems in any platform of EA, which require aligning all systems to the business vision, business requirements and information systems. Murray and Trefts (2000) state that:

The IT imperative is to construct enterprise-wide systems and capabilities needed by business to compete...EAs identify the means for collaboration between these different parts, in order to achieve the desired business objectives.

In this study, the shortcomings of this attribute are related to decisions, budget, business architecture and alignment. Outsourcing decisions, in regard to EA and its application, are made by technologists whereas business people, to whom application is applied, have less input to a decision.

The steering committee made the decisions based on recommendations made by technologists not from the business. And they certainly don't take a risk management approach to that. (Participant 1)

In addition, the technology budget depends heavily on its application. The higher the quality integration, the better the result of the application capability, but also is the higher the cost. Missing such a balance can affect other aspects. This means that the technical requirements and budget of an online application for twenty million people is higher than the technical requirements and budget of the same application for two million people.

When you have 300 or 400 systems in the backend and no integration, it's very hard to do online successfully.(Participant 9)

..more often it's not that we don't get scale or scalability. It's more often we know what should be but we only resource part of it.(Participant 5)

Moreover, business architecture may not be fully mature or implemented within the organisation and this certainly can affect its application, as explained by one participant: "We don't have fully populated business architecture. We are still struggling with that" (Participant 9)

Finally, alignment (as a subject matter) has become a dominant issue in all aspects of IT projects. In this regard, any misalignment of application and technology can have a direct impact on the anticipated benefits to be gained from a project. This view was expressed by one participant in response to a question about the challenges in presenting public values:

Yeah, the challenge is alignment, ensuring that our corporate capability is aligned to our benefits realisation and also to ensure that our plans at the local level are aligned to our plans at the corporate level. So, there's two sides to it.

(Participant 6)

Risk Assessment: The Risk Assessment attribute received the highest maturity level of Improved (4) in 7% of the fourteen projects and the lowest maturity level of Committed (2) in 43% of the projects. The remaining 50% of projects have an Established maturity level (3) (See also Figure 6-13 in Appendix J).

Although in this attribute only 57% of projects were at the Established or Improved maturity level, the attribute had a reasonable average maturity level of 2.7 across all projects. In addition, the attribute did not have an Ad Hoc maturity level (1) in any project. Various forms of risk control were taken, including compiling risks and issues in table format, Critical Success Factors (CSFs), risk estimation and value profiles, risk management plans and registering and risk mitigation. The findings indicate this attribute was important in the context of IT governance.

Five participants were asked 'How does organisation evaluate project risks?', and their answers are summarised in Table 6-2. Clearly, much attention has been paid to risk and a lot of risk mechanisms were adopted at various organisational levels and at an ongoing process. Their major purpose, as Table 6-2 indicates, is to identify those risks and mitigate them so that the anticipated value will be maximised.

TABLE 6-2: EVALUATION OF PROJECT RISK

Risk Mechanisms	Focus Areas	Purposes
Corporate Risk	National Risk Standards Community/political risk pressures	Minimising high level risk Minimising risk larger than project or program
ICT risk	CIO	Prioritising process risk calculation
Project risk, risk matrixes	Different part of organisation, project managers	Reviewing reports of steering committees Clearing any issues that require resolution Mitigating risk to Australian Standards
An approach to risk (risk methodology)	Project/program managers	Risk assessment Aligning to corporate strategy
Fifteen different categories of risk	Life cycle of projects where risks can be identified and quantified	Creating toolsets for Enterprise project management Mitigations are worked out
Risk workshops	Various stakeholders	Glean more on scope Intricacies of each project Detail of risk assessment

While an overall IT security awareness for project success in public organisations is supported by literature (Cresswell 2004; Cresswell, Burke et al. 2006; Gist and Langley 2007; NASCIO 2007), the findings show a negative relation between the Risk Assessment attribute and success rate. The negative contribution of this attribute to project success is not surprising, as during this case study the Council was undergoing a maturity process in many areas and that the Risk attribute itself suffered from its own well-being. For example, in Project 4, a rapid resumption of organisational services was the only solution available to mitigate the risks, which included food, bushfire, earthquake, terrorism, disease and tsunami. In Project 2 internal audits, training, testing and SLA were delayed because the software did initially work.

Moreover, project risk assessment is supposed to be evaluated based on a corporate risk management framework or based on a national risk management framework. In practice, that does not occur. The project management office which runs the ICT projects uses it own risk evaluation framework, even though that is based on staff's expectation of good consequences.

The identification of risk and quantification of risk is really done on ad hoc basis depending on the experience and knowledge and the people involved in doing in the business case. So it's relatively ad hoc and unstructured even though it looks like it follows risk standards, but it does not. (Participant 1)

Thus, assessing something and doing something about it can be different. Risk management in this organisation still requires more efforts in achieving the desired outcomes. While this attribute may be important, like the Training attribute, it failed to promote IT project success. The status of risk and its management, along with a call for improvements, were expressed by a number of participants.

Risk.Emm...This is a problem. Risk is basically evaluated by different people differently... So Risk at the moment is that an area we need too to address. The corporate risk area has had a couple of managers of managing corporate risk with where supposedly to go. So the corporate risk really needs to drive that our standard corporate risk process. And that it hasn't happened well to date.

(Participant 8)

..so yeah, so it's a pretty elementary risk analysis and, yeah, it will be ineffective backed up by whole bunch of stuff and actually performed by the office of the CIO that group takes on that role.(Participant 7)

The experience and knowledge of and the people involved in risk are crucial and hidden elements that improve risk mechanisms and associated management. If risk is properly managed and controlled, it will leverage benefits within the organisation.

Benefits to Organisation: The Benefits to Organisation attribute received the highest maturity level of Improved (4) in 29% of the fourteen projects. The majority level of this attribute is Established (3) in 50% of the projects. The only other maturity level is that of Committed (2) in 21% of the projects.

In the past, group officers set up the tasks needed to deliver projects. They placed limits themselves, using milestones to deliver a project on time and within budget. However, most key changes in ICT strategy over the last five years that have had an impact on organisations are related to a need to go beyond traditional methods of measuring project success, such as whether the project is on time and within budget, and to measure success in terms of the value delivered by the project. These benefits do not happen by themselves, but they can be delivered if they are properly defined. The focus on benefits realisation has been the biggest change in the organisation.

People are realising you really need to be putting in a system, you really need transition activities and organisational change needs to be properly managed in order to leverage the benefits within the organisation. (Participant 8)

In this study, this attribute was related to the design of internal benefits' identification in the business case and more details in project management plan. Identifying project benefits and milestones facilitates the development of a roadmap showing how such benefits can be measured and, thus, achieved.

Yeah. So, we are getting much better at managing the project life cycle and so once we get that down then I think we are going to improve the quality of the solutions we are delivering. Yeah, that's still an area for improvement.

(Participant 9)

The findings indicate that this attribute shows a positive relation to the success of projects due to the fact that staff are encouraged to achieve benefits by ensuring that all pieces of a project are strategically aligned to corporate plans and the corporation's future vision. Based on participant comments, the approaches of staff are improving. The importance of identifying benefits at an early phase for the project's sake is supported by the literature review (Foley 2006; Dodd, Yu et al. 2009; Mocnic 2011).

The case study shows a shifting of focus to a more public value culture in the organisation through an improvement of internal services and also through a focus on business change rather than on IT outcome.

So just making that switch from seeing them as ICT projects to seeing them as business projects with an ICT capability support is that we need to very much say this is a business project. These are the outcomes and benefits to the business and to the community...(Participant 10)

Benefits to Public: The Benefits to Public attribute received the highest maturity level of Improved (4) in 7% of the fourteen projects (Project 3). The predominant maturity level of this attribute was the Established maturity level (3), found in 64% of the projects, followed by the Committed maturity level (2) in 29% of the projects.

In this study, the Benefits to Public attribute has public-related features and was designed for public organisations that focus on customer services and the public's value of IT investments. The predominant maturity level of this attribute was either the Established (3) or Improved (4) maturity level, found in 71% of the projects. The findings reveal that the Benefits to Public attribute is important to a local government. Such an attribute shows to contribute positively to success across projects. However, if such benefits are not identified to be measured, this attribute can be the greatest inhibitor to obtaining benefits.

I suppose the most enabler one would be decision-making and the most inhibitor was the value realisation. (Participant 9)

Benefits were defined to be in the form of efficiency, effectiveness, political return or safer community and similar terms. "The vast majority, I would say, 95% of our outcomes plus are around community value." (Participant 10)

Due to the nature and complexity of a public organisation, quantification of benefits in financial terms has been a challenge as predicted by literature (ISACA 2007). This challenge also occurs in our case study organisation.

We are still very poor in a value realisation model, very poor... It's that because most of our benefits are really about efficiency and effectiveness and they are not particularly bankable...(Participant 9)

The findings were also in agreement with one of the participants:

Those things are difficult. Social and environmental are difficult to quantify but very important to our business and a lot of time the compelling argument without a full ROI will win maybe true.(Participant 1)

That participant's responses matched that in previous reports. Gerry Wethington, Missouri CIO, as cited by Cresswell (2004), states:

It [ROI] will let us begin to make assessments and decisions about funding a project or developing a new service based upon some true data. That moves you from having emotional debates about projects to having factual discussions.

Thus, the issues pertain to community values and may not be of economic benefit. On that basis, IT projects in this public organisation may be projects that are not cost effective (they are not making a profit). Therefore, the complexity of determining public value makes evaluating such benefits a difficult task. Such complexity, however, still needs to be addressed, as mentioned by one of participants.

..just because we are public it doesn't mean we are not accountable financially because we absolutely are. At the end of the day we have to demonstrate return on investment. We have to demonstrate net present value. We have to demonstrate costs through its full life cycle. (Participant 4)

Economic/Financial Metrics: The Economic/Financial Metrics attribute received the highest maturity level of Committed (2) in 86% of the fourteen projects and the lowest maturity level of Ad Hoc (1) in 14% of the projects. As with the Clarity of Direction and Align Technical Solutions with Business Solutions attributes, this attribute also has only two levels of maturity, but they are lower. For example, Clarity of Direction has maturity levels 3 and 4, Aligned Technical/Business Solutions has levels 2 and 3 and this attribute has levels 1 and 2. The average maturity level of this attribute across all projects is 1.9, the lowest average maturity level among the assessed attributes.

The Economic/Financial Metrics attribute has not strongly contributed to the success rate. The low maturity results of this attribute could be attributed to the nature of social and public benefits in the assessed projects in this organisation. Table 6-3 presents the major themes of the Economic/Financial Metrics attribute that were used in the organisation. The status mark (+ or –) refers to the strengths or weaknesses of each metric. The study reveals that there are number of economic and financial mechanisms already in place and implemented in the organisation. To some extent, as Table 6-3 indicates, some of them provide a good advantage to the organisation and "..they are effective in doing what they are supposed to do, which is perhaps to just deliver the project"(Participant 8).

However, in other situations these metrics have not resulted in advantage to the organisation. This result is due to the nature or type of benefits such as political and social return, which naturally are difficult to quantify. Another reason could be due to the nature of the rapid change of adopted systems which makes it difficult to compare measurements between old and new systems.

Benefits from IT investments are often difficult to describe and estimate directly in monetary terms.   (European Commission DG Enterprise 2003)

Being a public organisation that provides services, the low maturity of the Economic/Financial Metrics attribute was tolerated, as there were fewer financial measurement tools needed.

IT projects in the public will be running projects that aren't cost-effective, they won't make a profit. Because the role of a public provider is to provide services that economically may not be...(Participant 3)

Details of participants' observations on these metrics are given in Table 6-3

TABLE 6-3: THE STRENGTHS AND WEAKNESSES OF METRICS USED BY COUNCIL

Themes on economic/financial metrics	Status	Example quote
The usefulness of RIO	+	I think they are a good indicator of organisational performance in certain contexts. I think that again in government organisation it's more than the bottom line. It's about social and environmental returns. (Participant 1)
Mechanisms used	−	..we've got a lot of scope to improve our economic return of investment calculations and measurement because at the moment we do have waste because we don't do that part of the business well enough. (Participant 1)
	+	We thought about benefits process, now is establishing baseline measures. So for the tangible stuff obviously we try to at a point in time would sort of take a measurement. (Participant 11)
	−	..we have to be careful that however we measure that now in the old system, once we have changed whatever we are going to change, we have to be able to re-measure that in the future. So that's quite an issue. (Participant 11)
	−	In terms of the intangible stuff, probably we have done a few things like we have got baselines on staff surveys and stuff like that, but we probably haven't done enough delivery in that space to know that when we get to the end of benefits realisation whether we are going to be able to prove it or not.... Yes, difficult one. (Participant 11)
The effectiveness of metrics	±	On the whole they are effective in doing what they are supposed to do, which is perhaps to just deliver the project. Effective in doing that but not very effective at capturing and managing the change required to leverage the benefit from the project. (Participant 8)
The effectiveness of tools to measure the value of IT like BVIT	+	It works well...probably four...It does the job...You can always improve...People may want to improve it, but it's allowing us to prioritise our portfolio and in the three or so years that we have had it implemented no one has actually asked to change it(Participant 3)
	−	No...Value is perceived rather than quantifiable and it's usually spoken in terms of narrow technological outputs of a project as documented in the project management plan rather than organisational outcomes in terms of change. (Participant 1)
Clear achievement Line	+	...you are probably seeing, we have benefits, obviously each project has deliverables and timelines and cost, it also has benefits and probably the earlier projects were a bit sketchy on the benefits, but the later ones have got benefit tables in with names who is measuring and reporting it and so it's a lot clearer. In terms of benefit deliverables, I think we are hitting at the moment about 59% over the last two years in terms of achieving our benefits. 59% benefit delivery. (Participant 10)
Difficulty in measuring benefits	−	Sometimes, I don't think it's that hard to measure the benefits to the community I mean most of our services have their own benefit to the community or an outcome for the community. And most of them are able to be measured, we currently don't measure them, it doesn't mean to say we couldn't.... We could. The only thing that's required there was an investment in measurement....(Participant 10)

6.4 OVERALL ATTRIBUTES MATURITY SUMMARY: ACROSS-PROJECTS ANALYSIS

Attributes across projects: Table 6-1 presents the total maturity summary of all fourteen projects. In this chapter, SA attributes maturity levels were presented and analysed across the fourteen projects. Overall, the Clarity of Direction attribute received the highest average maturity level of 3.4 across projects, followed by the IT Investments and Budget and Benefits to Organisation attributes, which both received a 3.1 average maturity level. In contrast, the Economic/Financial Metrics attribute receives the lowest average maturity level of 1.9, followed by an average of 2.0 for the Training attribute.

Attributes within projects: As Table 6-1 indicates, the attribute maturity levels range between 1 and 4. Three projects (Projects 2, 8 and 13) have an Improved maturity level (4) in five of their attributes. Likewise, other three projects (Projects 1, 6 and 9) receive an Ad Hoc maturity level (1) in two of their attributes. Among all attributes, Clarity of Direction receives the highest maturity level of Improved (4) in five projects, followed by Benefits to Organisation, which receives the same maturity in four projects. On the other hand, Training received the lowest maturity (1) in five projects followed by Economic/Financial Metrics, which received a 2 maturity level in all projects except two (Projects 3 and 9) which have the lower Ad Hoc maturity level (1). The attribute maturity levels and their relations to success rate are summarised below

6.5 SA ATTRIBUTES MATURITY SUMMARY: ACROSS-PROJECTS ANALYSIS

In Section 6.3, the findings of an across-projects analysis of SA maturity and projects success were presented. That section aimed to address the research question RQ(b) *'Which Strategic Alignment attributes promote the success of IT projects?'*. This section summarises the findings in Section 6.3. The findings reveal that each of the attributes are important to the organisation, but the degree of their importance to project success differ. Nine attributes have a positive relation to the success rate and they have significant impact on project success. Four attributes have a moderate relation to the success rate, thus they have a relative impact on project success. The remaining two other attributes have a negative relation to success rate because they have a weak or no impact on project success. The last two attributes, Training and Risk, are not incorporated with literature, revealing the areas that need to be questioned. Figure 6-3summarises these findings.

FIGURE 6-3: SUMMARY OF RELATIONSHIP BETWEEN ATTRIBUTES AND SUCCESS RATE

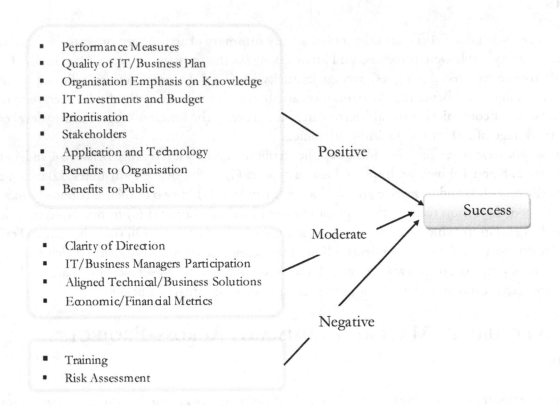

- Performance Measures
- Quality of IT/Business Plan
- Organisation Emphasis on Knowledge
- IT Investments and Budget
- Prioritisation
- Stakeholders
- Application and Technology
- Benefits to Organisation
- Benefits to Public

Positive

- Clarity of Direction
- IT/Business Managers Participation
- Aligned Technical/Business Solutions
- Economic/Financial Metrics

Moderate

Success

- Training
- Risk Assessment

Negative

7. Analysis of SA perspectives Maturity Levels Across Projects

7.1 General Remarks

Chapter 7 presents the overall results of the across-projects analysis of SA perspectives' maturity levels and the success rates of the projects. This chapter consists of six sections. Section 7.2 presents the overall maturity levels of the five perspectives. Section 7.3 presents the SA perspectives' maturity levels and their impact on success rate across projects. The analysis is aimed at addressing RQ(c): *'Which SA perspectives promote success of IT projects?'*. Section 7.4 presents an overall summary of SA perspectives maturity and the projects' success rate. Section 7.5 presents other patterns related to the projects. Section 7.6 concludes this chapter.

7.2 The Overall Perspectives' Maturity Levels

The results in Table 7-1 indicate that the perspectives' maturity levels range from a high of 4 to a low of 1.3. Among all perspectives, the Strategy perspective is the leading perspective in terms of maturity, and is the only perspective with an Improved maturity level (4) in Project 8. The Decision-Making perspective received the second highest maturity level of 3.7 in three of the fourteen projects (Projects 2, 13 and 14) while the Strategy perspective received the same maturity level in one project (Project 7). On the other hand, the Knowledge perspective has the lowest maturity level of 1.3 in Project 6, followed by a maturity level of 1.7 in Projects 1 and 5.

Table 7-1: Perspective Maturity Levels

Projects	Strategy	Knowledge	Decision-Making	Enterprise Architecture	Public Value	Average Maturity	Success Rate
Project 1	2.3	1.7	3	3	2.3	2.4	0.54
Project 2	3.3	2.3	3.7	2.7	3	3	0.88
Project 3	2.7	2.7	2.7	2.7	2.3	2.6	0.58
Project 4	2.3	2.3	2.7	2.3	3	2.5	0.68
Project 5	2.7	1.7	2.7	3	2.3	2.5	0.64
Project 6	2.3	1.3	2.7	2	2.3	2	0.46
Project 7	3.7	2.7	2.7	2.7	2.7	2.9	0.55
Project 8	4	2.3	3.3	3	2.7	3.1	0.96
Project 9	2.7	2	2.7	2.3	2	2.3	0.66
Project 10	3	2.7	2.5	3	2.3	2.7	0.34
Project 11	3	2.7	2.7	3	2.7	2.8	0.65
Project 12	2.7	2.7	3	2.3	2.7	2.7	0.63
Project 13	3	3	3.7	2.3	3	3	0.76
Project 14	3.3	2.3	3.7	3	2.7	3	0.75
Total Average	2.9	2.3	3.0	2.7	2.6		

Decision-Making and Strategy: As shown in Table 7.1, two perspectives (Decision-Making and Strategy) have the highest average maturity level of 3 and 2.9, respectively, across all projects. Those perspectives' maturity levels may reflect their importance, especially when the vision and budgetary aspects of the organisation are concerned.

Enterprise Architecture (EA) and Public Value (PV): Two perspectives, EA and PV, have a total average maturity of 2.7 and 2.6 respectively

Knowledge: The Knowledge perspective received the lowest average maturity level of 2.3 across all projects among the five perspectives. This maturity level accounts for 46 per cent of the average maturity level across the projects. This suggests that the area of knowledge perspective needs to be addressed in this organisation.

Details on the perspectives' maturity levels and their impact on success rate are provided in the following section.

7.3 PERSPECTIVES' MATURITY LEVELS AND SUCCESS RATES: ACROSS-PROJECTS ANALYSIS

7.3.1 STRATEGY PERSPECTIVE MATURITY AND ITS SUCCESS RATE ACROSS PROJECTS

Strategy (S): The Strategy perspective received the highest maturity level of Improved (4) in Project 8 and the lowest maturity level of Committed (2.3) in three projects; (Projects 1, 4 and 6). While those three projects are considered earlier projects, in that they took place in the first years of the case study, the results show that the maturity level of the Strategy perspective is higher in all seven later projects, with no project receiving less than a 2.7 maturity level. The findings indicate the growing maturity of this perspective over time. This perspective received the second highest average maturity level (2.9), after that of the Decision-Making perspective (3).

Previously we were walking along the path looking down on our feet. And now we are walking with eyes up to where we are actually going... Now we can start to navigate the path through on where we want to be and the CIO [is] driving us through. The CIO [is] communicating [with] that message what needs to be changed. The CIO rationally issued some communications that are very clear about what organisation is trying to achieve and they are measurable which hasn't actually been the case in the past...(Participant 5)

In comparison to the other perspectives, the Strategy perspective had the highest overall level of maturity, receiving at least an Established maturity level (3) in 50% of the fourteen projects, followed by the DM and EA perspectives, which both received maturity levels of Established (3) in 43% of the projects, then PV in 21% of projects and, finally, Knowledge perspective in 7% of projects. The findings indicate the dominance of the maturity of the Strategy perspectives among all of the perspectives.

I think within all government departments or government entities, there is a strong emphasis on strategic planning and they do very, very good strategic planning about this, that link from strategy to reality is that.(Participant 6)

The maturity level for this perspective showed consistent improvement over the five years of this case study. The organisation has a strong commitment towards its vision and strategy, and the findings indicate that this perspective has a strong impact on project success. Strategy is influential to the success rate.

The purpose of the Information Communications and Technology (ICT) Strategy is to provide the foundation for information and technology-related investments to 2009 to ensure resources are targeted to the areas that will deliver the greatest business benefit.(Council 2006, p.7)

The findings are consistent with the literature, where the importance of the Strategy perspective on successful project management is beginning to emerge (Grant 2003; Brown and Grant 2005; Slaughter, Levine et al. 2006; Oh and Pinsonneault 2007; Dodd, Yu et al. 2009). Moreover, this perspective is relevant to building a conceptual framework as it identifies the deficiencies areas where improvement is required (Green and Ali 2007; Al-Hatmi and Hales 2010).

Figure 7-1 shows the positive relation between the Strategy perspective and success rate. If the maturity of a project increases, the success of projects also increases.

FIGURE 7-1: STRATEGY AND SUCCESS RATE

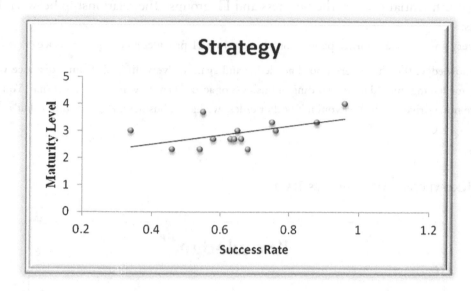

7.3.2 KNOWLEDGE PERSPECTIVE MATURITY AND ITS SUCCESS RATE ACROSS PROJECTS

Knowledge (K): The Knowledge perspective received the highest average maturity level of Established (3) in Project 13 and the lowest maturity level of Ad Hoc (1.3) in Project 6. As in the Strategy perspective, there is also an improvement in the maturity level of projects between earlier and later projects in the Knowledge perspective. For example, the maturity levels of Projects 1, 5 and 6 are 1.7, 1.7 and 1.3 respectively; they are all considered to have low maturity levels and all are earlier projects. On the other hand, the lowest average maturity level in later projects is Committed (2) in Project 9. This pattern of temporal improvement in maturity levels was not clear in the DM, EA and PV perspectives. This indicates that the Strategy and Knowledge perspectives have improved their maturity levels over time due to organisational changes and decisions that impact on those perspectives. The Knowledge perspective's improvement is credited to the establishment of the position of Chief Information Officer in the organisation.

Since the CIO came in, we have architectural layers. We are now able to make value-based decisions. We now have strategies. Not only are they developed but they are executed and delivered, which is just brilliant, and we are putting mechanisms in place for knowledge management and knowledge-sharing via centralised systems and repositories so everybody can share that information. So, I can quite happily say we are about a 3. I go to CIO summits annually in Sydney where I hear about national organisations and global organisations and I listen to them and I think "Oh goodness, you are where we were three years ago". So when I go to these things and compare us I can say relatively I can say that we are a 3.(Participant 4)

It should also be noted that the Knowledge perspective is considered to be the most important perspective after Strategy in terms of project success according to the participant interviews.

I think the first important one is Knowledge, it's No.1 because without knowledge you can't develop strategy or do anything else anyway. So for me Knowledge is No.1...And then No.2, I think is Strategy.(Participant 4)

While the efforts and effective communication channels being applied successfully in the organisation to improve the sharing of knowledge among staff, Figure 7-2, however, indicates this improvement has only had a moderate influence on project success. The findings show that with more communication there is better understanding and increased interaction between business and IT managers is required (Broadbent and Weill 1997; Elpez and Fink 2006;De Haes 2007). The Knowledge perspective was also affected by the Training attribute. As mentioned earlier in Section 6.3, the Knowledge perspective has some major limitations in relation to two influential groups, the business and IT groups. The relationship between business and IT was described as:

Pretty poor, very poor… The business people don't believe that IT provides a very good service.(Participant 9)

In terms of knowledge, they have very poor knowledge and again it's very difficult from a distance very hard for me to know and it struck me having limited understanding of business of accounts or how ICT can assist that. You think again maybe it's people. ICT are not a driver itself. We wouldn't bother unless we have a business. I'm not sure if that's always been seen as important.

(Participant 7)

FIGURE 7-2: KNOWLEDGE AND SUCCESS RATE

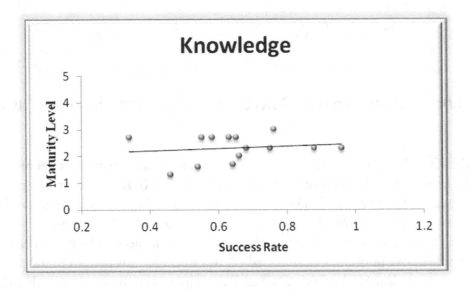

This finding is crucial and very important to public decision-makers. This is because the Knowledge perspective deals with staff and their skills. It is important that the level of effective communication, training and sharing knowledge remains at high level within staff to cope with the rapid changes of technology that enhance IT implementation. Therefore, the Knowledge perspective is an important area in which the organisation needs improvement. The reference model consists of characteristics (attributes) of Knowledge that need to be assessed and tracked, such as IT/Business Managers' Participation and Organisation Emphasis on Knowledge and Training.

7.3.3 DECISION-MAKING PERSPECTIVE MATURITY AND ITS SUCCESS RATE ACROSS PROJECTS

Decision-Making (DM): The Decision-Making perspective received the highest average maturity level of 3.7 in three projects (Projects 2, 13 and 14) and the lowest average maturity level of 2.5 in Project 10. This perspective, however, received the highest total average maturity level of Established (3) across the fourteen projects, followed by the Strategy perspective (2.9).

This perspective is found to be strongly influential on the success of IT projects, mostly as an enabler (Figure 7-7):

Okay, I suppose the most enabler one would be decision-making and the most inhibitor was the value realisation.(Participant 9)

The study shows that decisions on IT investments and budget, on prioritisation of projects and IT value and on the stakeholders involved (the business, IT and society representatives)were made by an IT governance structure group. This group consists of ELT, CGC, OCIO, ICT Portfolio Management Committee, BSRG and TRG (see Section 4.6.1). Those sections all worked together to produce a systematic approach to decision rights (Weill and Ross 2004). A systematic approach to decision-making on IT investments and budgets, prioritisation and stakeholders facilitates the implementation of IT projects. In addition, from the across-projects analysis, since the study also shows this perspective receives the highest average maturity level (3) among the five perspectives, the overall contribution of the systematic approach to decision-making was of benefit to the organisation.

FIGURE 7-3: DECISION-MAKING AND SUCCESS RATE

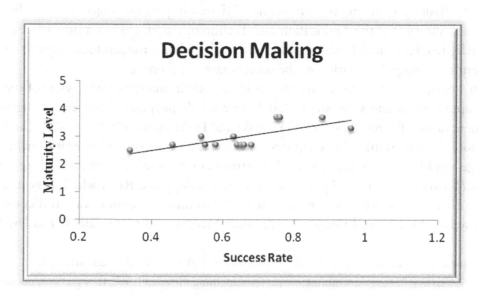

The study shows there were situations where decision delays produced a dollar cost. For example, Project 10 suffered some shortcomings in the decision-making aspects during implementation of the project, for example, the Steering Committee's inability to make timely decisions in unplanned situations, its procrastination in ending the project, its requests for additional information which required lengthy research and analysis and its requests for information that was outside the scope of the project (Council 2009). As a result, while high variation occurred in time as well as scope (functionality) and benefits expected, the variation in budget alone was estimated to be AU$3,796,450 over the original budget amount (Council 2009).

The challenge the organisation faces in this perspective is related to its decisions being dominated by technologists.

..there is a governance framework where the stakeholders are involved but the decision-making is dominated by technologists. Again, the decision is made very much from a technological mindset rather than business mindsets. (Participant 1)

7.3.4 ENTERPRISE ARCHITECTURE PERSPECTIVE MATURITY AND ITS SUCCESS RATE ACROSS PROJECTS

*Enterprise Architecture (EA):*The Enterprise Architecture perspective received the highest average maturity level of Established (3) in six projects (Projects 1, 5, 8, 10, 11 and 14) and the lowest average maturity level of Committed (2) in Project 6. The Enterprise Architecture perspective receives an average maturity level of 2.7 across all projects. The average maturity level in the seven earlier projects is 2.6 compared to 2.7 in the later ones, thus it has shown a slight improvement in maturity levels over time. While this perspective received an average 2.7 maturity level across all projects, some participants believe a more realistic figure would be around a maturity level of 2.

Enterprise architecture they've certainly got it so it's a... erm... That's tricky, because 3 is said 'integrated across the organisation'. Now, it's not, so I'm guessing it must transactional level, so I'm guessing level 2 will be it.

(Participant 7)

Enterprise architecture is probably not really my place to comment because it's not one of my core areas, but I would say it's more down a little bit at this level – 2.(Participant 11)

As with the Knowledge perspective, this perspective has a moderate impact on project success (Figure 7-4). The results might be influenced by its three attributes, Aligned Technical/Business Solutions, Application and Technology and Risk Assessment, as each one has different impact on project success. For example, the figure in Appendix K shows that the Application and Technology attribute has a positive relation to success rate whereas the Aligned Technical/Business Solutions attribute has a moderate relation to success rate and the Risk assessment has a negative relation to the success rate of IT projects.

Examining this perspective was necessary to provide a holistic enterprise-wide view of the organisation, consisting of the applications and subsystems that form a whole purposeful process of alignment between business requirements and information systems (Ylimaki and Halttunen 2005/06, p.189). The EA perspective is a useful visual tool for communication purposes. It provides a big picture of what it is included in EA and can be used to urge a public organisation to take EA attributes into consideration, including the attributes of Aligned Technical/Business Solutions, Application and Technology and Risk, which may identify the parts of an organisation that are essential for the purpose of controlling the complexity and consistent changes of information systems (Martin and Gregor 2002; Hjort-Madsen and Gotze 2004; Kaisler, Armour et al. 2005).

Like the domination of technologists in many decisions, EA is also affected similarly.

The minute we start saying, "Hang on a minute, we prefer something different here", they go "No, no, no, that doesn't fit the architecture". Pull a piece of paper and run away. So, yeah, we don't have much input or insight into how the IT architecture decision is made.(Participant 1)

FIGURE 7-4: ENTERPRISE ARCHITECTURE AND SUCCESS RATE

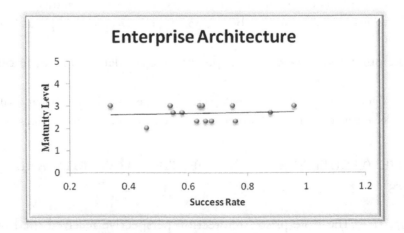

7.3.5 Public Value Perspective Maturity and its Success Rate Across Projects

*Pubic Value (PV):*The Pubic Value perspective received the highest average maturity level of Established (3) in three projects (Projects 2, 4 and 13) and the lowest average maturity level of Committed (2) in Project 9. The PV perspective hasan average maturity level of 2.6 across all fourteen projects. This perspective has not shown improvement over time, even though some participants believed that it had.

..probably the earlier projects were a bit sketchy on the benefits, but the later ones have got benefit tables in with names [of] who is measuring and reporting it, and so it's a lot clearer. (Participant 10)

Additional steps were added to corporate plans and performance measures and other value tools were considered, such as a safer community and capability of eGovernment initiatives to ensure the success of projects and the deliverance of the benefits expected.

But we are actually getting better now. We have actually got our benefits realisation plan now. It goes up with the business cases. And they live beyond the end of the life of the project…So after the project shuts down it goes back to him or her. And they have to deliver the benefits.(Participant 9)

Additionally, among the five perspectives, the Public Value perspective is considered the most important one in the organisation, as explained by one participant.

Public value is the most important as defined by the corporate plan, because public value is a reflection of the bold future vision and how we need to progress towards it in the next five years. So, I mean it is going to be that as a main driver of value for an institution such as us.(Participant 8)

However, that opinion contradicts the view suggested earlier (Section 7.3.2), that Knowledge is the most important perspective followed by Strategy. In fact, another participant mentions that the two perspectives are equally important to the organisation. The different rankings given to the importance of these perspectives indicates that different people perceive the perspectives differently because they see value benefits differently.

The study reveals the strong role this perspective plays in the success of IT projects (Figure 7-5). This appears to contribute to the fact that the Public Value attribute was considered as an important issue for the organisation and that importance was seen at the top management level. The research literature suggests that an alignment strategy for information technology governance (Grembergen 2004), the public ROI(Pardo and Dadayan 2006), alignment with public strategic goals (Firth, Mellor et al. 2008) and increased value for customers (Mocnic 2011) are all important to be considered in any proper public value plan. Without PV, evaluating and measuring success as well achieving benefits will be impossible.

FIGURE 7-5: PUBLIC VALUE AND SUCCESS RATE

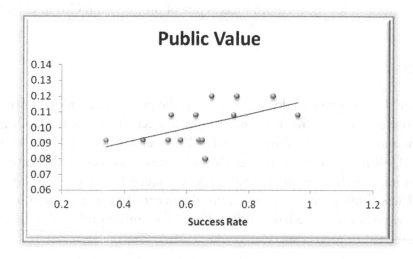

7.4 SA Perspectives Summary: Across-Projects Analysis

The maturity levels of the SA perspectives across the fourteen projects were presented in Section 7.2. The perspectives' maturity levels were compared to the success rate of the projects. A summary of those findings is presented in this section. As Figure 7-6 indicates, among the five perspectives, the results reveal that the S, DM and PV perspectives have a positive relation to the success rate. The Knowledge and EA perspectives have a moderate relation to the success rate. The findings imply that the Strategy, Decision-Making and Public Value perspectives are more significant to project success than the other two perspectives. However, other study such as Gregor, Hart and Martin (2007) has examined 1600 projects and found that Knowledge Management and Enterprise Architecture has been shown to have positive contributions to success. Therefore, it shall be noted that the low significance of later perspectives to project success in our study might be due to limited small sample size, 14 projects as its basis. In other words, while I am analysing 'large scale' constructs such as strategy, knowledge and public value, I am doing so using a limited sample.

In next section, other patterns (project characteristics) are explored, such as size, duration and documentation of projects and their relation to the success outcome.

FIGURE 7-6: SUMMARY OF PERSPECTIVES' RELATIONSHIP WITH SUCCESS RATE

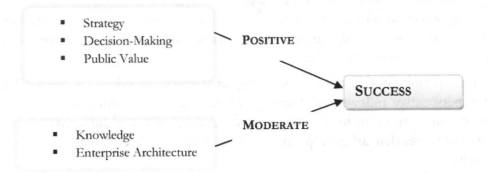

7.5 Other Patterns Observed Related to Projects

In Chapter 4 (Section 4.3.3), a case summary of project characteristics was presented. All projects were characterised according to the projects' size, duration time, documentation and project type (See also Table 4-1). These characteristics are then examined in detail in relation to the success outcome. The project type characteristic is excluded as it has only one project type, IT. In this section, Group A stands for the lowest success rate of projects, Group B for the medium success rate of projects, and Group C for the highest success rate of the projects.

7.5.1 Project Size

Does project size matter? In other words, does having a specific project size group contribute to success rate projects? The projects undertaken are characterised as large projects [more than AU$2m], medium projects M(a) [between AU$1m and AU$2m], medium projects M(b) [between AU $500,000 - AU$1m] and small projects [under AU$500,000]. Figure 7-7 illustrates this relationship as seen in this case study. The findings reveal that projects that have medium or highest success rate projects (in Group B or C) have at least 50% of a similar project size. These findings, however, are contradictory to the other group (Group A), which shows projects that have the lowest success rate have, in fact, 75% of a similar project size. The project size in this study, therefore, is not related to project success

FIGURE 7-7: PROJECT SIZE

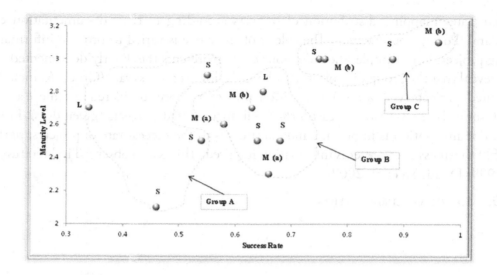

7.5.2 PROJECT DURATION

Figure 7-8 illustrates project duration distribution, from short duration (less than six months) to long duration (more than two years). The duration of the most projects was medium duration (six months) to long duration (two years). The findings reveal that projects that have the highest success rate (see Group C) have at least 75% of similar duration of Medium (a), from six months to a year. Similarly, the projects that have the medium success rate (Group B) have at least 50% of projects with a similar duration. This finding indicates that the projects that have similar duration facilitate the project success rate, regardless if they are short, medium or long duration.

FIGURE 7-8: PROJECT DURATION[22]

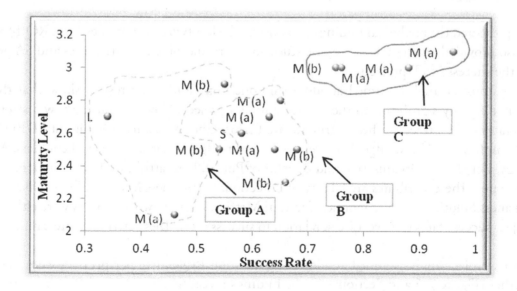

---

22   L: Long duration (+2 yrs),   M (a): Medium duration (6moths- 1 yr)

M (b): Medium duration (1 yr – 2 yr)   S: Short duration (under 6 months)

7.5.3 PROJECT DOCUMENTATION

Figure 7-9 illustrates mapping distribution of projects according to their documentation characteristics; 'Well', 'Less', and 'Poorly documented'. The colour of the circle is varied to provide information about the corresponding project. For example, the red colour circle represents the 'Poorly documented' of the project. The findings reveal that the group of projects that have the lowest success rate (Group A) include at least 75% 'Less documented' projects (yellow dots). In addition, projects that are considered to achieve a medium or high success rate (Groups B and C) have at least 50% 'Well documented' projects (green dots). From this we can conclude that documentation is important and contributes to the success rate of projects, particularly when the quality of IT/business plan presents information required. This is also observed in literature (Morgan and Schiemann 1999; Dodd, Yu et al. 2009).

FIGURE 7-9: PROJECT DOCUMENTATION

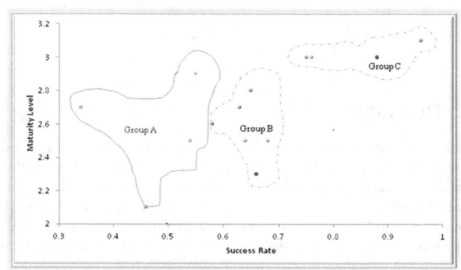

7.6 CONCLUSION

Chapter 7 has presented the analytical findings that answer the last two research questions: RQ(b) and RQ(c). The chapter also provided an analysis of the case study to determine which SA attributes and SA perspectives contribute to the success of IT projects.

In the across-project analysis aimed at addressing question RQ(b), the results show that the assessed attributes are not equally significant to the success rate of a project. Nine attributes show a strong relation to the success rates of IT projects. These attributes are Performance Measures, Quality of IT/Business Plan, Organisation Emphasis on Knowledge, IT/Investments and Budget, Prioritisation, Stakeholders, Application and Technology, Benefits to Organisation and Benefits of Public. Four attributes have a moderate relation to project success rates. These attributes are Clarity of Direction, IT/Business Managers' Participation, Aligned Technical/Business Solutions and Economic/Financial Metrics. Finally, two attributes were either negatively related to project success rates or have a weak relation to success rates. These attributes are Training and Risk Assessment.

In the across-project analysis aimed at addressing question RQ(c), five perspectives were examined. As mentioned earlier (Figure 7-6 and Section 7.4), the findings reveal that S, DM and PV have strong impacts on project success rates compared to the impact of other perspectives such as K and EA perspectives, which both show a moderate impact level on the success rate of IT projects.

The implications of these findings and further discussions of the issues are presented in the next chapter.

8. DISCUSSION AND CONCLUSION

8.1 GENERAL REMARKS

A review of the literature concerning issues surrounding IT project success (Section 2.5) revealed that ensuring the success of IT projects has become a challenging task for the public sector. The high expenditures related to technology and the increasing number of failures of IT projects in many countries are important issues and they suggest the need for more proactive involvement in the control of IT activities (ISACA 2007). The high accountability associated with decision-making and collaboration among executives and stakeholders together with a systematic approach are now considered to be at the core of obtaining success of an IT project (Aurum, Wohlin et al. 2006; NASCIO 2007). This study, therefore, sought to investigate the impact of strategic alignment (SA) on the success of IT projects in a local government over a five-year period.

This chapter summarises the issues addressed in this research. Section 8.2 revisits the literature review in brief, particularly concerning IT projects in public sector organisations. Section 8.3 evaluates the research model and highlights the main findings of the research. Section 8.4 provides the contribution of the research to practitioners in order to maximise the value of government IT initiatives and reduce the failure rate of IT projects by implementing guidance identified in the holistic conceptual model that examines maturity levels of SA perspectives and attributes and their projects outcomes. Comments on the implications for public policy-makers and public practices are given in details in Section 8.5, which is followed by Section 8.6, which describes the limitations of the study. Future trends and a final conclusion are given in the last section, Section 8.7.

8.2 REVISITING THE LITERATURE REVIEW IN BRIEF: IT PROJECTS IN PUBLIC ORGANISATIONS

Governing information technology (IT) is difficult and is recognised as a critical issue facing the public sector today. IT itself delivers no value without careful planning to achieve the potential of IT alignment with business goals. This concern is a focus of many governments today (Chan and Reich 2007; De Haes and Grembergen 2008; Raven 2008; Dodd, Yu et al. 2009; Mocnic 2010; Grabski, Leech et al. 2011; Lobur 2011).

While some public sector organisations are able to deliver their IT projects on time and on budget, few they actually identify and measure the benefits supposedly arising from those projects (Gershon 2009). The criteria of delivering on time and on budget are no longer sufficient to judge the success of IT projects for public organisations. Other criteria such as the scope and value of projects are more important to consider. In other words, a project may be executed correctly, come in on time and be within budget, but fail to realise any acceptable benefits. In the case study, Project 12 showed this tendency (Council 2009). This appears to be due to a combination of reasons, including not strategically aligning IT with government vision in the initial preplanning phase of the project.

The leadership uncertainty in Project 10 was another issue (which had a documentation characteristic of 'well documented') where the vendor failed to deliver the promised capability during project implementation.

Moreover, as in many organisations, defining the success of IT projects has been an issue to the public organisation assessed here.

Thus, if what is considered 'success' in a project is not clearly defined, the success measurement will also suffer and affect the project outcome. Such findings corroborate the findings from other studies. For example Krignsman (2011)notes:

> Difficulty defining IT success is one reason that failure statistics are all over the map. For example, when I compiled CRM failure to starts from various sources, for 2001 to 2009, the numbers ranged from 18 per cent to 70 per cent.

Thus, both program and project managers were required to partner with customers, understand organisational drivers and care enough to ensure that the project they lead delivered the business results for which it was designed. Those issues are part of what is considered project success.

On the other hand, other participants (Participant 1 and 9) do not think that the desired benefits have been significantly achieved in the organisation due to the fact that the project managers who are doing projects were allowed to evaluate themselves, whether or not they did a good job, they generally said they did a good job which is not so right.

Thus achieving public IT value has proven to be a challenge to public organisations (Lobur 2011). This is caused by many factors including:

- Being able to measure time and budget factors, but fail to measure anticipated benefits (Gershon 2009);
- Internal capability of leadership to decision-making was uncertainty in some cases (projects) (Dos Santos and Reinhard 2010; Heiskanen 2012);
- Defining success has not always been an easy task where different stakeholders should be taken into account during planning process to define how to measure tangible and intangible public benefits (Lee et al. 2011; Merkhofer 2011; Krigsman 2011); and
- Communicating between program/project managers and between business/IT managers was not effective enough to solve problems they encountered (Tremblay 2006; Muniz 2009).

The influence of different perspectives on stakeholders contributes to enriching the understanding of how project success is defined in the organisation. Based on the case study, four success factors have been identified that form a basis for describing project success. These factors are budget, time, scope and value realisation. Together, and with a certain degree of tolerance, these factors contribute to a meaning of project success that can be monitored and measured. This study indicates that success rate varies from 0.34 to 0.96 (see Table 5-3, Section 5.4). Based on data obtained from Section 5.5, the overall success rate has shown a clear improvement over time, as was observed in the maturity level. Overall, the study does show that the organisation enjoys a higher project success rate in later projects.

One of the major findings of this study is that simply introducing formal standards and mechanisms into the process cannot guarantee the success of an IT project. It is important to include SA perspectives into a holistic approach that address the issues of the IT implementation process from a concept plan and business case to delivering value of IT at the end. The process that contributes to the success of IT projects matters.

8.3 RESEARCH MODEL REVISITED

Based on the review of the literature, five SA perspectives were identified as contributing to the success of IT projects. These perspectives are Strategy, Knowledge, Decision-Making, Enterprise Architecture and, Public Value. In this regard, a reference model was proposed that would achieve better linkage between business and IT strategies and with the purpose of generating public value through IT projects (Figure 2-5). The research model was evaluated through a case study of IT projects conducted in a public sector organisation. The research proposal was to find out whether SA maturity of projects has an impact on the project's success rate. The findings from RQ(a)[23] reveal that the impact of SA maturity level on project success is positive. The findings from this study are confirmed in existing research literature relating to the impact of strategic alignment on IT project success (Byrd, Lewis et al. 2005; Cresswell, Burke et al. 2006; Elpez and Fink 2006).

Figure 8-1 portrays an emergent research model that illustrates the impact of strategic alignment perspectives and attributes on the IT projects success along with the sub research questions. The comprehensive model developed is simple but effective, consisting of a maturity level of strategic alignment perspectives and implemented criteria called attributes for each of the perspectives. This contribution advances an understanding of how government policy-makers can optimise the value of their IT projects, framing the general contribution of this research.

---

23   What is the relationship between the SA maturity level and the success rate of IT projects?

FIGURE 8-1: EMERGENT RESEARCH MODEL

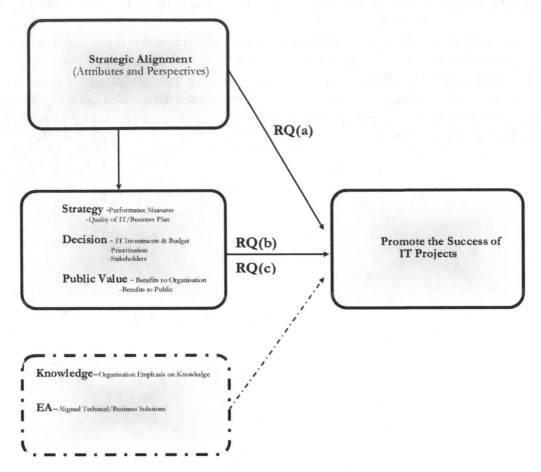

Based on RQ(b)[24] and RQ(c)[25], the study further identified SA perspectives and attributes that positively contribute to the successful implementation of IT projects.

As Figure 8-1 indicates, only nine attributes out of the fifteen attributes examined could be shown to have had a positive impact on success rate of projects. These attributes are Performance Measures, Quality of IT/Business Plan, Organisation Emphasis on Knowledge, IT/Investments and Budget, Prioritisation, Stakeholders, Application and Technology, Benefits to Organisation and Benefits to Public.

Similarly, only three perspectives (out of the five perspectives examined) were shown to have a strong positive impact on the success rate of the projects. These perspectives are Strategy, Decision-Making and Public Value.

The key findings from this study are:

- A successful alignment between business and IT strategies is evident when both IT and business strategies can demonstrate a planned alliance which then leads to tangible and successful business-focused outcomes (Gartlan and Shanks 2007). This case study supports the view that emphasising isolated components of ICT strategy without links to corporate objectives can be misleading. Local government, with its emphasis on SA, seems to perform better as an organisation with improved

---

24   RQ (b): *Which Strategic Alignment attributes promote the success of IT projects?*

25   RQ (c): Which Strategic Alignment perspectives promote the success of IT projects?

K and EA perspectives have not shown a strong evidence to contribute to the project success and as such present avenues for future research that may provide a comprehensive view of SA in public sector

levels of IT project delivery. Moreover, putting formal structures and processes in place does not guarantee success. By examining SA perspectives and attributes with respect to the IT strategy, public value, management portfolio and corporate activity, public decision-makers can improve the success of government IT projects.

- SA perspectives can occur anywhere in the process. Hence, a holistic approach to SA perspectives was proposed (particularly those identified in the model such as Strategy, Decision-Making and Public Value and their attributes) to improve the government's ability to assess public returns on IT investments. The criteria presented in this model identify attributes that affect each SA maturity level. An implication from this study is that the absence of these attributes will contribute to the low SA maturity level, which affects project implementation. Examining a local government in Australia provides an opportunity to better understand the impact of alignment perspectives on IT projects in public organisations. Such understanding can be used to as guide to other public organisations. Hence, identification of the importance of implementing SA perspectives within a government agency is one of the main contributions of this book.
- The study shows that the term 'success' in relation to IT projects was inconsistently defined by stakeholders. As a result, many projects suffer when project participants are in disagreement as to the proper success emphasis or goals for the project (Griffith and Gibson Jr. 2001).

The study offers four factors to measure success (Time, Budget, Scope and Value Realisation) that have been previously suggested in the literature. Crawford (2011), ITGI (2008), Ambler (2007) and McNamara (2005) identify their ranking weight and their degree of tolerance (Krigsman 2011) as success measurements. But, as the case study shows, without an agreement on what the organisation wants to achieve and how to measure it, the organisation will always get an open-ended discussion about the degree of success that the organisation has to achieve. "By not aligning measurements and rewards, you often get what you are not looking for" (Jack Welch, CEO of General Electric Corporation, 2001 as cited by Middleton (2004, p.387).

- Sharing knowledge and responsibility with an Executive Leadership Team (ELT), the Corporate Governance Committee, the ICT Portfolio Management Committee, the Office of the Chief Information Officer and the Technical Representative and Business Solution Representative Groups is a key element of SA. However, in a local government other relevant stakeholders, such as vendors, the Customer Account Representative and the Chief Executive Officer, should also be involved. Their participation would improve the effects of context and planning behaviours on end-users and vice versa. However, engaging stakeholders such as suppliers or vendors should be done with caution, as they are kind of trying to sell their products.

The improvement in the Knowledge perspective results indicate that the knowledge and effective communication associated with knowledge, are seen as 'power' and as a 'strategic tool' to the success of an organisation (Martinette and Dunford 2004; Tremblay 2006; Wang and Belardo 2009). The average maturity level of this perspective (2.3), however, is still far lower compared to the other perspectives.

The quote above illustrates that public organisations are often characterised by the need for collaboration, yet collaboration among staff does not always arise spontaneously. However, collaboration among IT and business managers is needed to increase the chance of project success. Tremblay (2006, p.1) states

Interestingly, while some segments of the public sector are embracing knowledge management, our experience is that a series of unfounded perceptions are contributing to the delay of its wider adoption.

Unlike the Strategy and Knowledge perspectives, which both show improvement in their maturity through time, other remaining perspectives, Decision-Making, Enterprise Architecture and Public Value, show overall stable maturity over time.

The study also shows that an organisation can implement a range of intellectual assets, initiatives such as data, systems, information, staff leanings and experiences, sharing knowledge and policies and, finally, the use of information and knowledge management mechanisms. These initiatives maximise benefits and minimise risks and costs, based on public services and community expectations, by providing direction on how an organisation manages the information and knowledge assets. This evolves engaging stakeholders in the decision-making process on public ICT investments.

In this study, the attribute Organisation Emphasis on Knowledge shows a positive contribution to the project's success. The findings are consistent with concept of the power of knowledge that has been suggested by many academics and practitioners (Broadbent and Weill 1997; Carr, Folliard et al. 1999; Elpez and Fink 2006; Firth, Mellor et al. 2008; Raven 2008; Wang and Belardo 2009), and encouraged by Kearn and Sabherwal (2007) as well as Gartlan and Shanks (2007), as a means to build a robust SA theory in IS.

- 'On time and on budget' is no longer considered an indication of the success of IT projects. In a recent review of the Federal Government, Gershon (2009) found that while 52 per cent of government IT projects were delivered on or under budget and 43 per cent were delivered within timeframes, only 5 per cent reported that they actually identified and measured the benefits from a project. In this study, the results reveal that it is often not until the end of the project that the organisation examines what they were supposed to achieve, by which time it is too late to make any changes that will improve the success, although top executives knew the scope of projects and adopted formal project management methodologies. Public agencies, by their nature, are often more complex, have more stakeholders and are less concerned about financial issues than they are with private values which reflect their social, economic and political priorities. Therefore using time and budget only as primary criteria is relevant, but inadequate to measure benefits for public organisation.

8.4 CONTRIBUTION OF THE RESEARCH

The increasing number of failures of IT projects in public agencies indicates that significant issues remain unsolved(Burke and Cresswell 2006; NASCIO 2006;Lobur 2011). Many public agencies still face difficulties in delivering their IT projects on budget, on time and within scope in order to achieve the value expected from IT investments (Vogt and Hales 2010). Hence, SA issues were examined from various perspectives found in the literature and an effective methodology (the reference model) was developed. Here, one of the main contributions was in the examination of five perspectives and fifteen attributes and their relation with the success rate of IT projects. The first conclusion is that benefits from IT investments accrue if SA perspectives are deployed in effective IT governance context. Three SA perspectives, Strategy, Decision-Making, and Public Value were found to be most significant to the project success in this organisation.

Another contribution of this research is to expand the existing research into an area that helps determine why IT projects still fail. This need has been expressed by academics and practitioners for many years(Williams 2005; Weerakkody, Janssen et al. 2007). Therefore, this study was designed to answer the main research question: *'How are SA perspectives deployed in IT governance context to ensure government IT projects' success?'* The study closely examined how this deployment has affected the success of IT projects. On time, on budget and within scope are all crucial factors, but are insufficient criteria for determining the success of IT projects. This study indicates that having a value realisation criterion is important and is a crucial element of measuring project success. There is no value to an IT project when both budget and scope are on track but value realisation remains unknown. The findings also indicate that some projects are doomed to failure due to issues including misalignment and poorly defined objectives.

The third contribution of this study is that it enhances the understanding of the antecedents of SA perspectives by examining the relationship between attributes' maturity and project outcomes in terms of time, budget, scope and value realisation. The implication of the study is that the higher the maturity levels of the attributes, the better the outcome. A significant finding revealed some interesting insights about the need for taking care of the issue of SA perspectives throughout the design as well as the issue of IT governance mechanisms over the organisation's IT resources and the issue of operating effectively if the organisation is to achieve its objectives (Green and Ali 2007). This will help decision-makers in public agencies to focus on those attributes and perspectives of alignment that contribute to success of their IT projects.

The fourth contribution of this study is that it has identified perspectives and attributes that strongly, moderately and negatively affect the success of IT projects. The study provides clear evidence of misalignment in areas where significant deficiencies of perspectives and attributes were identified. The findings also reveal that in this organisation some attributes, such as Organisation Vision, Budget and Benefits to Organisation are given more attention by management team than others. This attention, however, does not necessarily contribute positively to the success rate. For example, the Application and Technology attribute (2.8 maturity level) contributed positively and more highly to the success rate than the Clarity of Direction' attribute, which had a higher maturity level (3.4). An implication of this result is that a good balance of limited resources should be assigned to those attributes that contribute more to the success of IT projects. The findings correspond to the general notion found in research literature (Sledgianowski and Luftman 2005; Lange, Lee et al. 2011) where project prioritisation reduces costs by generating maximum value for the resources available, eliminates the need of endless meetings and enhances the achievement of the benefits anticipated as explained by Merkhofer (2011, p.1):

> Resources are insufficient to do everything at once. You need to understand the costs, risks, and benefits of each project and decide which to fund now, which to postpone, and which to scale back or eliminate.

A further contribution of the study is that achieving benefits (or value realisation) is hard to measure, particularly in a public organisation where the quality of customer services, political returns and other forms of IT value are counted. Though the objectives and goals of most of the projects in the organisation were clear, the study indicates using traditional economic and financial metrics measurements does not present an accurate picture of success. The findings reflect the nature of a public agency that focuses on customer and citizen services rather than on financial returns.

A final contribution of the study is related to the holistic conceptual model proposal that would facilitate the alignment of various perspectives and attributes in IT projects and their life cycle. The model assists business and IT managers in an organisation to increase the success of their projects. The conceptual model provides further research into SA aspects and perspectives in a government context. The research model, when applied to a local government, confirmed that SA issues that were already addressed had contributed to IT project success.

8.5 IMPLICATIONS FOR PUBLIC POLICY-MAKERS AND PRACTICES

A number of key implications for public policy-makers and public policy practices arise from this study, and are described below.

8.5.1 CREATING PARTNERSHIP ALLEGIANCE AMONG STAKEHOLDERS TO AVOID IT PROJECT FAILURE

Public organisations are characterised by having many stakeholders. These stakeholders have different views and perspectives about projects, IT values, methodologies applied, length of experience, area of expertise and their experts' levels of authority and decision; thus their input and participation within projects are important to shape the way projects are defined and monitored to maximise benefits and minimise risks. This can be enhanced by effective communication, training, sharing knowledge (knowledge is also known as being empowering to staff) and encouraging the participation of external stakeholders and community representatives. This in turn will create a long-term partnership allegiance among stakeholders that enhances appropriate public policies on ICT investments, which in turn leads to quality service via successful IT project implementation. The notion of a large number of stakeholders in a public organisation implies that immense efforts from IT governance management are needed, along with considerable care, to create a mutual understanding among stakeholders as a vehicle to promote best practice in project management.

8.5.2 UNDERSTANDING THE LEVEL OF IMPACT OF SA PERSPECTIVES AND ATTRIBUTES ON IT PROJECTS

This paper showed that SA plays a significant role in IS, based on the descriptive data analysis of IT projects. The actual relationship between SA perspectives' and SA attributes' maturity levels on one hand and the success rate of IT projects on the other hand is evidence of this tendency. The findings in Section 5.5 reveal that the higher the SA maturity, the better the result. The findings indicate that SA perspectives and attributes are both important and both should be integrated in the planning process for the better results of IT investments. The implication of these findings is that the positive attitudes of these perspectives and attributes contribute to harmonising and easing the implementation of IT projects. Individual perspectives and attributes may or may not contribute to the success of IT projects' implementation (as discussed in Sections 6.4 and 7.3), but, since strategic alignment perspectives can occur anywhere in the process, an implication from this study is that it is essential to consider the perspectives that can be deployed in order to practically contribute to the success of IT projects, regardless of the standards and mechanisms that are already available. If these perspectives and attributes have not been included in a holistic approach to organisation, their contributions will suffer. This suggests that IT projects are liable to fail if an organisation is not aligned. In his study of the importance of organisational alignment to facilitate information systems adoption, Peter Middleton(2004) reported that only 48 per cent of staff questioned believed that the existing structure did assist change; this low figure again points towards a poorly aligned organisation. An obvious example of poor alignment in this case study is where the ICT strategy of the organisation was well articulated and communicated at the top level but, at the same level, failed to demonstrate an effective communication between business/IT participation in some projects investigated.

For example, COBIT, a necessary framework, measures the process maturity in a local government, but it is an inadequate tool to measure how the capability is used and deployed. Instead, the conceptual SA model was utilised in this research to provide improved visibility and possible solutions to potential problems encountered in Council's present structures. By using the conceptual model, various aspects of IT projects and Council business can be improved. The overall strategy, archetypes and mechanisms of IT governance are all harmonised in such a way that the overall direction can move smoothly towards desirable IT behaviours. In addition, the study provides guidance on how public sectors can reduce their redundant IT investments and lower the rate of IT project failures by implementing SA perspectives throughout their whole organisational process. This contribution advances the understanding of government policy-makers of how public sectors can optimise the value of their IT projects.

8.5.3 CREATING A HIGH-QUALITY PROJECT PLAN BEFORE IMPLEMENTATION

The study demonstrated that the methodology used in Business Cases, Project Management Plans and Project Charters can be very useful to the creation of a business/IT plan, especially in an organisation that has a mature IT governance and project management approach. Reich and Benbasat (1996) suggested that the degree in which a high-quality set of interrelated IT and business plans exists in organisation is important. One of the key aspects of project approval was the need to provide adequate and accurate information about the proposed projects, when the anticipated cost/benefits analysis includes the proper use of the benefits measurement methodology.

Having a high-quality business/IT plan seems to facilitate the implementation of project and strengthen positive outcomes where resources, milestones, technical specifications, business and functionality requirements, risks, deliverable, constraints/dependencies and related issues of alignment and priority to organisation strategy are well defined, addressed and explained in detail. The implications of this finding, in relation to the success of IT projects, is that this information provides a very useful means to encourage and promote proper project management practices. The study also indicates what may be expected from project outcomes and what may be expected from different stakeholders in a public organisation. Expectations that will play a role in shaping the way projects are monitored, benefits are maximised and risks are minimised.

8.6 LIMITATIONS OF THE STUDY

The result of this study should be considered in light of a number of limitations. However, some of these limitations are seen as fruitful avenues for future research under the same theme.

The scope of the research was limited to a single-case design in a local government and was based on observation, interviews and government documents. The issue of generalising findings to new situations is a weakness inherent in the reliance on a single case study and looms larger in this type of study than in other types of qualitative research. Hamel(1993, p.23) observes that "the case study has basically been faulted for its lack of representativeness".

While it is not the intention of this case study to provide generalisable findings in the conventional qualitative sense, it was an objective to provide a case description that would be theoretically useful to predicting the success of IT projects in similar settings, but without actually conducting such a study (Stake 2005). Much can be learned from this particular case, especially those aspects that have some application to the broader context. Erickson (1986) argues that since the general lies within the particular, what is learned in a particular case can be transferred to similar situations.

Another limitation of the study is that this study used theoretical reasons (theoretical sampling) to select cases and participants which, as a result, limited the generalisation applicability of the findings. In addition, extreme views or less lightly weighted views were not removed from analysis due to a limited number of projects and participants and thus self selection bias might exist. To minimise the biased nature of the sample, various procedures were put in place to manage these potential problems (see Section 4.3.2). In addition, a wider spectrum of participants, IT projects and organisation characteristics across organisation were obtained to ensure that the cases and participants involved in the study were the most appropriate representatives of those involved in IT projects management within public organisation.

This research is based on interviews and participant observation, an approach that is mostly appropriate with small populations and with accessible activities and frequent events over a certain period. However, the researcher was considered as an 'outsider' by the participants (Miles and Huberman 1994); hence, placing limitations on the findings. This is because the participants are uncertain as to what extent they can protect the self-interests that are evident in their responses. Moreover, the semi-structured and open-ended interviews used in this study allowed the participants a straightforward way to discuss issues that they felt comfortable talking about. Given the sensitive nature of this social issue in the workplace and to help to avoid the difficulties associated with people's awareness of being participants in research, a number of procedures were conducted (Section 3.6), including the obtaining of confidentiality agreements and ensuring that the researcher's intentions were clear. In addition, the wide variety of documents and sources were gathered that reflect the organisation's everyday operations helped to obtain the data that was needed. Finally, inclusion of informal follow-up meetings was designed to develop a greater rapport with participants and, thus, overcome any reluctance to moderate responses.

While it is believed that all methods of collecting data were considered and that those selected worked well, the possibility of the presence of single case study bias and researcher bias cannot be ignored. Hence, those sources of bias should be kept in mind when making use of these findings.

Despite these limitations, the researcher has suggested a conceptual template, data coding and analysis approach that may be useful for further reviews of case study research (Gibbert and Ruigrok 2010). The study provides an opportunity to assess the impact of SA perspectives and attributes in a natural series of events and investigates the issues of projects outcome. Hence, the research may provide a meaningful and coherent picture of a single local government setting. While this study adopted a qualitative approach based on case study methodology, the findings should be useful as a basis for future quantitative studies.

8.7 RECOMMENDATION FOR FUTURE STUDIES

The problem investigated in this research concerned factors influencing the success of IT projects. These factors are the so-called Strategic Alignment (SA) Perspectives. Based on a literature review and on organisational structure and knowledge theories, a reference model of SA Perspectives was developed to better achieve alignment between business and IT strategies.

A significant implication of the study is that the delivery of IT projects on time and on budget is an inadequate measurement for value realisation from IT projects. A consideration of alignment domains is essential and this can be employed to increase the understanding needed for better achievement of value realisation in a public organisation. If these domains are put into a holistic approach and IT governance is implemented properly (De Haes and Grembergen 2008), many limitations in public organisations should be overcome. It is crucial to investigate the role of SA in the success of IT projects. However, despite the wide-ranging recognition of the importance of strategic alignment in IS, research on achieving and sustaining such alignment is lacking (Hirschheim and Sabherwal 2001; Bhide 2008; Estrin 2008). This may be explained by the evidence of so many different attempts, methods and models of measuring the returns on IT investments. This study concludes that the concept of SA was, and remains, a complex system consisting of interdependent subsystems that produce a purposeful whole (Peterson 2001).

In order for public organisations to prosper and maintain their IT project success, they must make effective use of their key resources and align those resources with their business strategies. This is especially true when dealing with decisions and when sharing knowledge among stakeholders in a public organisation. Knowledge, unfortunately, is one of the most valuable resources that organisations possess but it has been largely ignored in the projects reviewed and it received the lowest average maturity level of 2.3.

The study creates a basis on which further research into the SA of project success can be undertaken. This study investigates the impacts of SA perspectives and attributes on IT projects in a single local government in one period (2004–2009). Future research could investigate these impacts on IT projects within a single government sector but during two different periods. A key aspect of this further investigation would be, for example, to compare the SA maturity levels of both older and more recent projects and to determine how these differences can affect project outcomes. The findings of this study also provide the potential to conduct future studies across a larger number of projects, including in multi-government agencies. This could involve undertaking additional case studies of a larger number of public sector organisations.

The study highlighted a number of specific perspectives and attributes of SA that would be worthwhile including in further investigations, including the identification of those SA aspects in IT governance contexts, in project management and among personnel project managers that need to be better understood. Chief among these is the influence of the sharing of knowledge and decision-making among stakeholders and the influence of IT governance structure maturity over IT projects in public organisations.

A further opportunity for research would be to explore the influence of SA perspectives and attributes on different project types. This study focused on the impact of SA on IT projects only and only on four success factor criteria. Future studies may replicate this study but include different types of government projects to determine whether the use the same factors identified in this research will remain relevant to other types of projects

Finally, it is envisioned that numerous aspects of this study will be of considerable use in carrying out other research opportunities and it is hoped that this researcher will have the opportunity to pursue as least some of them in the future.

This concludes Chapter 8 and the book. The bibliography and appendices follow this chapter.

9. REFERENCES

AG (2004). Demand and value assessment methodology. I. M. Office. Canberra, Commonwealth of Australia. **5**: 115.

Al-Hatmi, A. and K. Hales (2010). Stategic alignment and ITproject in public sector organization: Challenges and solutions. European, Mediterranean and Middle Eastern Conference on Information Systems, Abu-Dhabi, UAE, ISeing.

Al-Hatmi, A. and K. Hales (2011). The impact of strategic alignment perspectives on IT projects: Analysis and discussions. Oman 2011 International Business Conference, SQU in Oman, SQU.

ALIA (2003). Federal Budget 2002. A. L. a. I. Association, ALIA Publishing, http://www.alia.org.au/publishing/budget. analysis/2002.html.

Allen, R. F. and C. Kraft (1987). The oganisational unconscious. NY, Morristown.

Ambler, S. W. (2007). "Defining Success." Retrieved 15 March, 2011.

Anisetty, P. and P. Young (2011). "Collaboration problems in conducting a group project in a software engineering course." Journal of Computer Sciences in Colleges **26**(5): 7.

Ann, B and John, D. (2012). 'Outlook', www.networkworld.com retrieved date 27 May 2012 P17-9

Aurum, A., C. Wohlin, et al. (2006). "Aligning software project decisions: A case study." International Journal of Software Engineering 16(6): 23.

Babbie, E. (2001). The practice of social research. USA, Thomson.

Baxi, M. and PSVillage, Eds. (2007). Tips from the treaches: The collective wisdom of over 100 professional services leaders, Santa Clara, CA: PSVillage Press, c2007.

Benbasat, I., D. K. Goldstein, et al. (1987). "The case reseach strategy in studies of information systems". Mis Quarterly **11**(3): 369-386.

Benbasat, I., D. K. Goldstein, et al. (1987). "The case research strategy in studies of information systems." MIS Quarterly **11**(3): 369-386.

Berger, P. L. and T. Luckmann (1967). The social construction of reality. New York, Doubleday.

Beynon-Davies, P. (2002). Information systems: An introduction to informatics in organisations. Basingstoke, Palgrave.

Bhansali, N. (2007). Strategic Alignment in Data Warehouses. School of Business Information Technology, RMIT University. Doctor of Philosophy: 395.

Bhatnagar, S. C. and N. Singh (2010). "Assessing the impact of E-Government: A study of project in India." USC Annenberg School for Communication adn Jounalism **6**(2): 109-127.

Bhide, A. (2008). The venturesome economy: How innovation sustains prosperity in a more connected world. USA, Princeton University Press, Preinceton, NJ.

Bittler, R. S. and G. Kreisman (2005). Gartner enterprise architecture process evolution.

Blaikie, N. (2000). Designing social research. UK and USA, Polity Press.

Bonoma, T. V. (1985). "Case research in marketing: opportunities, problems, and a process." Journal of Marketing Research **22**: 199-208.

Boyatzis, R. E. (1998). Transforming qualitative information: Thematic analysis and code development. United Kingdom, SEGA Publications, Inc.

Boynton, A. C. and R. W. Zmud (1987). "Information technology planning in the 1990's: directions for practice and research." MIS Quarterly 11(1): 59-71.

Brancheau, J.C., and Wetherbe, J.C. Key issues in information systems management. *MIS Quarterly, 11,* 1 (1987), 23–45.

Braun, C. and R. Winter (2007). Integration of IT Service Management into Enterprise Architecture. SAC '07. Seoul, Korea., @ 2007 ACM: 5.

Bridgman, P. and G. Davis (2004). The Australian Policy Handbook. Sydney, Allen & Urwin.

Broadbent, M. and P. Weill (1993). "Improving business and information strategy alignment: learning from the banking industry." IBM Systems Journal 32(1).

Broadbent, M. and P. Weill (1996). "Managing by maxim: Creating business driven information technology infrastructures." Sloan School of Management.

Broadbent, M. and P. Weill (1997). "Management by maxim: how business and IT managers can create IT infrastructures." Sloan Management Review 38(3): 77(16).

Brown, A., D. (1995). Organisational culture. London, Pitman.

Brown, A. E. and G. G. Grant (2005). "Framing the frameworks: A review of IT governance research." Communication of the Association for Information Systems 15: 696-712.

BSI (2011) Balance scorecard institute.

Buckley, J. W., M. H. Buckley, et al. (1976). Research Methodology and Business Decisions. New Your, National Association of Accountants (NAA).

Burke, G. B. and A. M. Cresswell (2006). The Austrian Federal Budgeting and Bookkeeping System. Public ROI - Advancing Return on Investment Analysis for Government IT: Case Study Series. NY, University at Albany, SUNY: 12.

Byrd, T. A., B. R. Lewis, et al. (2005). "The leveraging influence of strategic alignment on IT investment: An emprical examination." Information and Management 43: 13.

Canada, A. (2007). Millions wasted as tories eliminate gun registry contract, opposition says. Canada, The Auditor General of Canada.

Carr, D. C., K. A. Folliard, et al. (1999). "How to implement a successful communication program: A case study." Bell Labs Technical Journal: 8.

Carr, M. (1997) Risk management may not be for everyone, IEEE Software, Vol. 14, No. 3, pp. 21-24

Chan, Y. E., S. L. Huff, et al. (1997). "Business strategic orientation, information systems strategic orientation, and strategic alignment." Information Systems Research 8(2): 125-150.

Chan, Y. E. and B. H. Reich (2007). "IT alignment: an annotated bibliography." Journal of Information Technology 22: 316-396.

Chang, H., hsiao, H. et al. (2009). "Assessing IT-business alignment in service-oriented enterprises". Proceedings of the Pacific Asia Conference on Information Systems, Hyderabad, PACIS 2009.

Chen, Y., H. M. Chen, et al. (2007). "Electronic government implementation: A comparison between developed and developing countries. ." International Journal of Electronic Government Research 3(2): 45(17).

Cleland, D. and L. Ireland (2004). Project manager's portable handbook, McGraw-Hill, USA.

Cohen, J. (2003). A cross-national comparison of determinants and consequences of IS-business social alignment. the 4th Annual Global Information Technology Management World Conference, Calgary, Canada.

Collins, J. and J. I. Porras (2004). Built to last: Successful habits of visionary companies HarperBusiness.

Cooke-Davies, T. (2002). "The 'real' success factor on projects." International Jounal of Project Management **20**(3).

Corsten, D. and N. Kumar (2005). "Do suppliers benefit from collaborative relationships with large retailers? An empirical investigation of efficient consumer response adoption." Journal of Marketing 69(July): 80-94.

Council (2006). ICT strategy 2005-2009. Office of the Chief Infromation Officer. I. Services. Council, Queensland, Australia., @ 2006 GCCC. Version 1.0:52.

Council (2007). Strategy Framework, @ 2007 Council: 26.

Council (2007). ICT Project Management Framework. C. a. C. Governance, Council.

Council (2009). Project Closure Report. O. S. P. M. Office., @ 2009 Council: 24.

Counihan, A., P. Finnegan, et al. (2002). "Towards a framework for evaluating investments in data ware housing." Information Systems Journal **12**: 321-338.

Crawford, J. K. (2011). The strategic project office. U.S.A, CRC Press, Taylor & Francis Group.

Crawford, L. H. (2009). "Government and governance: The value of project management in the public sector." Project Management Journal **40**(1).

Cresswell, A. M. (2004). Retern on investment in information technology: A guide for managers. S. University at Albany. Albany, NY 12205, @ 2004 Center for Technology in Government: 46.

Cresswell, A. M. and G. B. Burke (2006). The Government of Israel's Merkava Project. Public ROI - Advancing Return on Investment Analysis for Government IT: Case Study Series. NY, University at Albany, SUNY: 14.

Cresswell, A. M. and G. B. Burke (2006). The Washington State Digital Archives. Public ROI - Advancing Return on Investment Analysis for Government IT: Case Study Series. NY, University at Albany, SUNY: 13.

Cresswell, A. M., G. B. Burke, et al. (2006). Advancing return on investment analysis for government IT: A public value framework. Center for Technology in Government. New York, University of Albany.

Creswell, J. W. (2009). Research design : Qualitative, quantitative, and mixed methods approaches, SAGE.

Crocker, L. and j. Algina (1986). Introduction to Classical and Modern Test Theory. Toronto, Wadsworth Publishing Company.

Croteau, A. M. and F. Bergeron (2001). "An information technology trilogy: Business strategy, technological deployment and organizational performance." Journal of Strategic Information Systems 20(2): 77-99.

Cuenca, L., A. Boza, et al. (2010). Enterprise engineering approach for business and IS/IT strategic alignment. 8th International Conferene of Medeling and Simulation - *MOSIM'*10. Hammamet - Tunisia, MOSIM' 10. **10**: 12.

Dailymail (2012), http://www.dailymail.co.uk/news/IT-project-failure-Labours-12bn-scheme-scrapped.html#ixzz1vJmFFS1T Saturday, May 19 2012

D'Angelo, A. and G. Abramo (2009) The alignment of public research supply and industry demand for effective technology transfer: The case of Italy. Science and Public Policy **36**, 13

Darke, P., G. Shanks, et al. (1998). "Successfully completing case study resarch: combining rigour, relevance and pragmatism." Information Systems Journal **8**(4): 16.

Davis, L., B. Dehning, et al. (2003). "Does the market recognize IT-enabled competitive advantage?" Information and Management **40**(7): 705-716.

Dawes, S., S., G. B. Burke, et al. (2006). The Commonwealth of Pennsylvania's Integrated Enterprise System. Public ROI - Advancing Return on Investment Analysis for Government IT: Case Study Series. NY, University at Albany, SUNY: 16.

De Haes, S. (2007). The impact of IT governance practices on business/IT alignment in the Belgian financial services sector. Management School. Bengin University of Antwerpen. **PhD:** 221.

De Haes, S. and W. V. Grembergen (2008). Analyzing the relationship between IT governance and business/IT alignment maturity. Proceedings of the 41st Hawaii International Conference on System Sciences, Hawaii.

De Souza, R., R. Nariwawa, et al. (2003). IT business spending to recover in Late 2003, Gartner.

Deloitte (2005). "Sustaining growth." 29 Sept., 2011.

DeLone, W. and E. R. McLean (2003). "The DeLone and McLean model of information systems success." Journal of Management Information Systems **19**(4): 9-30.

DeLone, W. H. and E. R. McLean (2002). Information systems success revisited. 35th Hawaii International Coference on System Sciences, Hawaii, 2000 IEEE.

Denzin, N. K. (1978). The research Act. New York, McGraw-Hill.

Denzin, N. K. and Y. S. Lincoln, Eds. (1994). Qualitative research. CA, US, Thousand Oaks, SAGE Publications, Inc.

Diener, E. and R. Crandall (1979). Ethics in Social and Behavioural Research, Univeristy of Chicage Press.

Dixon, P. and D. John (1989). "Technology issues facing corporate management in the 1990s." Mis Quarterly **13**(3): 247-255.

Dodd, S., P. Yu, et al. (2009). Successful project management: The implementation of strategy in local government. UK Academy for Information Systems, UK, AIS Electronic Library (AISeL).

Dos Santos & Reinhard (2010). 'Barriers to Government Interoperability frameworks adoption'. Proceeding AMCIS, AIS, p. 1-10 www.aisel.aisnet.org

Dowse, A. (2003). The benefits, limitations and governance implications of federated public sector systems. ACIS 200314th Australian Conference on Information Systems, Perth.

Duhan, S., M. Levy, et al. (2001). "Information systems strategies in knowledge-based SMEs: The role of core competencies." European Journal of Information Systems **10**: 25-40.

Dyer, W. G. (1986). Cultural change in family firms: Anticipation and managing business and family transitions. San Francisco, California, Josse, Bass.

Earl, M. J. (1993). "Experience in strategic information systems planning." MIS Quarterly **17**(1): 1-24.

Ebrahim, Z. and Z. Irani (2005). "E-government adoption: Architecture and barriers." Business Process Management Journal **11**(5).

Eddie, K. (2011),"Using training and development to recover failing projects", Human Resource Management International Digest, Vol. 19 Iss: 4 pp. 3 - 6

EDP, A. (2005). Business plan and future directions A. F. A. C. E. D. P. A. N.S.W, @ 2005 Australian Institute of Police

Management

EDP, A. (2007). Participation assegnments: Business Plan and Future Directions Papers, @ 2007 Australian Institute of Police Management.

Eisenhardt, K. M. (1989). "Building theories from case study research." The Academy of Management Review **14**(4): 18.

Eisenhardt, K. M. and M. E. Graebner (2007). "Theory building from cases: Opportunities and challenges." Academy of Management Jounal **2007**(1).

Eisenhardt, K. M. and C. B. Schoonhoven (1996). "Resource-based view of strategic alliance formation: Strategic and social effects in entrepreneurial firms." Organization Science **7**(2): 136-150.

El Sawy, O. A., A. Malhotra, et al. (1999). "IT-Intensive value innovation in the electronic economy: Insights from marshall industries." Mis Quarterly **23**(3): 305-335.

Elpez, I. and D. Fink (2006). "Information system success in the public sector: Stakeholders' perspectives and emerging alignment model." Information Science and Information Technology **3**: 12.

Erickson, F. (1986). Qualitative methods in research on teaching. Old Tappan, NJ., MacMillan.

Estrin, J. (2008). Closing the innovation gap: Reigniting the spark of creativity in a global e conomy. New York, McGraw-Hill.

European Commission DG Enterprise (2003). IDA Value of Investment. Sweden, European Commission DG Enterprise: 1-60.

Evans, N. (2004). "Promoting business/IT fusion: An OD perspective." Emerald **26**(4): 15.

Fardal, H. (2007) ICT strategy in an ICT user perspective: Exploring alignment between ICT users and managers. **4**, 12

FEA (2007). Value to the mission: FEA practice guidance. O. Federal Enterprise Architecture Program Management Office, www.whitehouse.gov. **5:** 56.

Feagin, J., R. , A. Orum, M., et al., Eds. (1991). A case for the case study, Chapel Hill, NC: University of North Carolina Press.

Fleiss, Joseph L. 1971. "Measuring nominal scale agreement among many raters." Psychological Bulletin 76(5):378-382.

Fielding, R. (2002). IT projects doomed to failure.

Firth, L. A., D. J. Mellor, et al. (2008) The negative impact on nurses of lack of alignment information systems with public hospital strategic goals. Australian Health Review **32**, 3

Fisher, J. (2001). "User satisfaction and system success: Considering the development team." Australian Journal of Information Systems **9**(1): 21-29.

Foley, K. (2006). Using the value measuring methodology to evaluate government initiatives.3rd Annual Global Crystall Ball User Cofference, Denver, www.anarisco.com.br/gerenciador/uploads/governamentais_2a.pdf.

Frankfort-Nachmias, C. and D. Nachmias (1996). Research Methods in the Social Sciences. United States of America, St. Martin's Press, Inc.

Galbraith, C. and D. Schendel (1983). "An empirical analysis of strategy types." Stratagic Management Journal **4**: 153-173.

Galliers, R. D. and D. E. Leidner, Eds. (2003). Strategic Information Management: Challenges and Strategies in Manging Information Systems, Syndetics Solutions, LLC.

Galup, S. D., R. Dattero, et al. (2007). "IT Service management: Academic Programs Face the Issues of Globalization." SIGMIS

CPR'07.

Garson, G. D. (1999). Information technology and computer applications in public adminstration: Issues and Trends. US, Idea Group Publishing.

Gartlan, J. and G. Shanks (2007). "The alignment of business and information technology strategy in Australia." Australia Journal of Information Systems **14**(2): 113-139.

Gartner (2004). www.gartner.com.

Gartner (2007).www.gartner.com.

Gartner (2011). "Top tends shaping the future of Australian data centers." Retrieved 4th, May 2011.

Gemini, C. (2000). M2PressWIRE.

Gershon, P. (2009). Money for something. The magazine of the Australian institute of project management. Australia, Australia College of Project Management (ACPM).

Gibbert, M. and W. Ruigrok (2010) The 'What' and 'How' of case study rigor: Three Strategies Based on Published Work. Organizational Research Methods **13**, 710-737

Gibbs, G. R., Ed. (2008). Analysing qualitative data. The Sage Qualitative Research Kit. London, SAGE.

Gibson, G. E. and M. R. Hamilton (1994) Analysis of pre-project planning effort and success variables for capital facility projects.

Gichoya, D. (2005). "Factor affecting the successful implementation of ICT projects in government." Journal of e-Government **3**(4): 175-184.

Gist, P. and D. Langley (2007). "Application of standard project management tools to research-A case study from a multi-national clinical trial." Journal of Research Administration **38**(2): 8.

Golafshani, N. (2003). Understanding reliability and Validity in Qualitative Research. The Qualitative Report. Toronto, Canada, University of Toronto. **8:** 10.

Golafshani, N. (3002). Understanding reliability and Validity in Qualitative Research. The Qualitative Report. Toronto, Canada, University of Toronto. **8:** 10.

Goode, W. J. and P. K. Hatt (1952). Methods in social research. New York, McGraw-Hill.

Government, G. W. (2009). "WiBe 2009- Economic efficiency assessments: WiBe Framework for ICT and eGovernment Projects." Retrieved 31/12/2009, 2009, from www.eu.wibe.de.

Grabski, S. V., S. A. Leech, et al. (2011). "A review of ERP research: A future agenda for accounting information systems." Journal of Information Systems **25**(1): 41.

Grant, G. G. (2003). "Strategic alignment and enterprise implementation: the case of Metalco." Journal of Information Technology **18**: 16.

Green, P. and S. Ali (2007). "IT governance mechanisms in public sector organization: An Australian context." Journal of Global Information Management **15**(4): 41(23).

Gregor, S. D. Hart, et al (2007) "Enterprise architectures: enablers of business strategy and IS/IT alignment in government", Information Technology & People, Vol. 20 Iss: 2, pp.96 - 120

Grembergen, W. V. (2004). Strategies for Information Technology Governance. London, @ 2004 Idea Group Inc.

Griffith, A. F. and G. E. Gibson Jr. (2001). "Alignment during preproject planning " Journal of Management in Engineering **17**(2): 7.

Gross, G. (2009), "Cloud computing, security to drive US Gov't IT spending", CIO, available at: www.cio.com (accessed January 10, 2010).

Guardian, T. (2008) Not fit for purpose: 2bl pounds cost of government's IT blunders.

Guldentops, E. (2004). Governaning infromation technology throught COBIT. Strategies for information technology governance. M. Khosrowpour and J. Travers. USA, Idea Group Publishing: 269-310.

Guldentops, E. (2011) Knowing the environment top five IT issues. IT governance

Guthrie, J. (1997). "Performance information and programme evaluation in the Australian public sector." International Journal of Public Sector Management **10**(3).

Gutierrez, A., J. Orozco, et al. (2006). Using tactical and organizational factor to assess strategic alignment on SME. European and Mediterranean Conference on Information Systems (EMICS). Costa Blanca, Alicante, Spain, ISEing.

Heiskanen, A (2012, p.1). Project Portfolio Management, https://www.isaca.org/ecommerce, ISACA, US. Retrieved on 26 may 2012.

Hales, K. R. (2005). Value creation in a virtual world. School of Information Technology. Gold Coast City Council, Bond University. **Doctor of Philosophy** 202.

Hamel, J. (1993). Case study methods: Qualitative research methods. CA, SAGE, Thousand Oaks.

Hannaway, C. and G. Hunt (1999). The Management Skills Book. England, Hants.

Hartung, s., B. H. Reich, et al. (2000). "Information Technology in the Canadian forces." Canadian Journal of Administrative Sciences **17**(4): 285-302.

Heeks, R. (2008). "ICT4D 2.0: The next phase of applying ICT for international development." @ 2008 IEEE **41**(6): 26.

Henderson, J. C. and N. Venkatraman (1992). Transforming organizations. Strategic alignment: A model for organizational transformation through information technology. T. A. Kochan and M. Useem, Oxford University Press US, 1992.

Henderson, J. C. and N. Venkatraman (1993). "Strategic alignment: leveraging information technology for transforming organization." IBM Systems Journal **32**(1): 4-16.

Hirschheim, R. and R. Sabherwal (2001). "Detours in the path toward strategic information systems alignment." California Management Reviews **44**(1): 87-108.

Hjort-Madsen, K. and J. Gotze (2004). Enterprise architecture in governance-towards a multi-level framework for managing IT in government. ECEG04, Dublin, Ireland, ECEG04.

Hodgkinson, S. L. (1996). The role of the corporate IT function in the federal IT organization New York, Oxford University Press.

Hoepfl, M. (1997). "Choosing qualitative research: A primer for technology education researcher." Journal of Technology Education **9**(1): 47-63.

Huang, W. W., J. D'Ambra, et al. (2002). Key factors influencing the adoption of e-government in Australian public sectors. Eighth Americas Conference on Information Systems (AMCIS), AIS Electronic Library (AISeL).

Huber, N. (2002). 'Gartner: Firms waste 251 bn pounds each year on III-conceived IT projects?'. UK, ComputerWeekly.com.

Hugoson, M.-A., T. Magoulas, et al. (2010). Enterprise architecture design principles and business-driven IT management. Business Information Systems Workshops. W. Abramowicz, R. Tolkdorf and K. Wecel. Verlag Berlin Heidelberg, German, @ 2010 Springer. **57:** 11.

Institute, I. G. (2006). Enterprise Value; governance of IT Investments, The Val IT Framework, IT Governance Institute, IL 60008 USA.

Institute, P. M. (2008). A guide to the project management body of knowledge (PMBOK), Project Management Institute.

Investopedia (2011). Investopedia, www.investopedia.com/dictionary.

Irava, W. J. (2009). Familiness Qualities, Entrepreneurial Orientation and Long-term Performance Advantage. Faculty of Business, Technology, & Sustainable Development. Gold Coast, Queensland, Bond University. **Doctor of Philosophy:** 260.

ISACA (2007). COBIT Case Studies, ISACA.

ITGI (2003) IT control objectives for Sarbanes-Oxley and board briefing on IT governance. .
ITGI, Ed. (2005). COBIT. Rolling Meadows, IL 60008, IT Governance Institute, USA.

ITGI (2005). IT Alignment: Who Is in Charge? USA, IT Governance Institute: 29.

ITGI (2006). "IT governance in practice : Insight from leading CIOs."

ITGI (2007). IT governance round table, ITGI.

ITGI (2007). Val IT Case Study: Value Governance- Police Case Study.

ITGI (2008). Enterprise value: Governance of IT investments. The Val IT framework. USA, IT Governance Institute, IL 60008 USA: 44.

ITGI (2008). IT governance global status report, IT Governance Institute.

Ittner, C. D. and D. F. Larcker (2001). "Assessing empirical research in managerial accounting: a value-based management perspective." Journal of Accounting and Economics **32**(1-3): 349-410.

Johnson, R. B. (1997). "Examining the validity structue of qualitative research." Education **118**(3): 10.

Johnson, S. D. (1995). Will our research hold up under scrutiny? Journal of Industrial Teacher Education. **32:** 3-6.

Johnson, S. D. (1995). Will our research hold up under scrutiny? Journal of Industrial Teacher Education. **32:** 3-6.

Johnston, H. R. and S. R. Carrico (1988). "Developing capabilities to use information strategically." MIS Quarterly **12**(1): 37-48.

Jones, D. (2009). M2 Presswire, @ 2009 Factiva: 2.

Joppe, M. (2000). "The Research Process." Retrieved Retrieved Feb 25, 1998.

Jung, Y. G. G. (1999). "Planning for computer integrated construction." Computer Civil Eng **13**(4): 217-225.

Kaisler, S. H., F. Armour, et al. (2005). Enterprise architeching: Critical problems. 38th Hawaii International Conference on System Sciences (HICSS'05) Hawaii, @ 2005 IEEE.

Kajalo, S., R. Rajala, et al. (2007). "Approaches to strategic alignment of business and information systems: A study on application service acquisitions." Journal of of Systems and Information Technology **9**(2): 155-166.

Kane, K. (2005). Managing change to reduce resistance. Boston, Harvard Business School Press.

Kaplan, R. and D. Norton (1992). "The Balance Scorecard: Measure that Drive Performance." Harvard Business Review **70**(1): 70-79.

Kaplan, R. S. and D. Norton (1992). "The balance scorecard: Measure that drive performance." Havard Business Review **70**(1): 70-79.

Kaplan, R. S. and D. Norton (2004). "Strategy maps." Harvard Business School.

Kaplan, R. S. and D. P. Norton (2001). "Transforming th balanced scorecard from performance measurement to strategic management: Part 1." Accounting Horizons **15**(1): 87-104.

Kearns, G. S. and R. Sabherwal (2007). "Strategic alignment between business and information technology: A knowledge-bases view of behaviors, outcome, and consequences." Journal of Management Information Systems **23**(3): 129-162.

Keen, P. G. W. (1991). Shaping the futurebusiness design through IT. Cambridge, MA, Harvard Business School Press.

Kerzner, H. (2009). A systems approach to planning, scheduling and controlling. Canada, John Wiley & Sons, Inc.

Khaiata, M. and I. Zualkernan (2009). "A simple instrument to measure IT-business alignment maturity." Information Systems Management **26**(2): 14.

Khosrow-Pour, M. (2000) Challenges of information technology management in the 21st century.

Khosrow-Pour, M., Ed. (2006). Cases on information technology: Lessons learned. Cases on Information Technology Series, Hershey, Ideal Group, c2006.

Khosrowpour, M., Ed. (2006). Cases on information technology: Lessons learned. Cases on Information Technology Series, Hershey, Ideal Group, c2006.

Klein, H. K. and M. D. Myers (1999). "A set of principles for conducting and evaluating interpretive field studies in information systems." Mis Quarterly **23**(1): 67-94.

Kotsik, B. and N. Tokareva (2007). UNESCO IITE contribution to e-inclusion policy development for education of students with disabilities. ICTA'07, Hammamet, Tunisia, ICTA' 07.

Krauth, J. (1999). "Introducing information technology in small and medium sized enterprises." Studies in Informatics and Control **8**(1).

Krigsman, M. (2011). "CIO analysis: Defining IT project 'success' and 'failure'." Retrieved 15th March 2011, 7th March 2011.

Krippendorff, K. (2003). "The dialogical reality of meaning." The American Journal of Semiotics **19**(1-4).

Lainhart, J. (2008). ISACA's New CGEIT credential meets business demands for IT governance. Certificate Magazine. **February 2008**.

Lange, D., P. M. Lee, et al. (2011). "Organisational reputation: A review." Journal of Management **37**(1): 49.

Lankhorst, M. M. (2004). "Enterprise architecture modelling: The issue of integration." Engineering Computing and Technology **18**(4).

Laudon, K. C. and J. P. Laudon (2004). Sistemas de Informacion Gerencial. Nueva edition, Prentice Hall.

Lederer, A. L. and a. L. Mendelow (1989). "Coordination of information systems plans with business plans." Journal of Management Information Systems **6**(2): 5-19.

Lederer, A. L. and H. Salmela (1996). "Toward a theory of strategic information systems planning." Journal of Strategic Information Systems **5**(3): 237-253.

Lee, O.-K., P. Banerjee, et al. (2006). "Aligning IT Compenents to Acheive Agility in Globally Distrubuted System Development." Communication of the ACM **49**(10): 49-54.

Lewis, R. I. (2006). Project Management. Project Management: 110.

Lin, C., G. Pervan, et al. (2005). "IS/IT investment evaluation and benefits realisation issues in Australia." Jounal of Research and Practice in Information Technology **37**(3).

Linchko, L., E. and A. J. K. Calhoun (2003). An evaluation of vernal pool creation projects in new England: Project documentation from 1991-2000. Environmental management. New York, Springer. **32**: 141-151.

Lincoln, Y. S. and E. Guba, G. (1985). Naturalistic Inguiry. USA, SEGA Publications, Inc.

Lobur, J. M. (2011) The success of a COTS caseload management system n State Government. Insights: Project Management

Locke, K. (1996). "Rewriting the Discovery of Grounded Theory After 25 Years?" *Journal of Management Inquiry* 5(3): 239-245.

Luftman, J. (2000). "Assessing business-IT alignment maturity." Communication of Association for Information Systems **4**(14).

Luftman, J. and T. Brier (1999). "Achieving and Sustaining Business-IT Alignment." California Management Review **42**(1): 109-122.

Luftman, J. N. (2009). Managing Information Technology Resources.

Luftman, J. N. (2011). Managing information technology resources: Leadership in the information age: 1339.

Luftman, J. N. and R. Kempaiah (2008). "Key issues for IT executives 2007." MIs Quarterly Executive **7**(2): 99-112.

Luftman, J. N., P. R. Lewis, et al. (1993). "Transforming the enterprise: The alignment of business and information technology strategies." IBM Systems Journal **32**(1).

Lynda M. Applegate, Robert D. Austin, et al. (2007). Corporate Information Strategy and Management. 1221 Avenue of the America, New York, NY, 10020., The McGraw-Hill Irwin Companies, Inc.

Macaulay, A. (2004) Enterprise architecture design and the integrated architecture framework.

Macquarie (2005). The Macquarie Dictionary. New South Wales, The Macquarie Library Pty, Ltd., Australia.

Maes, R. (1999). "Reconsidering information management through a generic framework." PrimaVera: 99-115.

Maes, R., D. Rijsenbrij, et al. (2000) Redefining business-IT alignment through a unified framework. Landelijk Architectuur Congres

Marchand, D. A., and Peppard, J. (2008), "Design to fail: Why IT projects underachieve and what to do about it", Granfield University, UK.

Marshall, C. and C. B. Rossman (1989). Designing Qualitative Research. United States of America, SEGA Publications.

Marshall, J. P. (2006), Information technology, disruption and disorder: Australian customs and IT, Proceeding conference The Association of Internet Research, Brisbane.

Martin, N. and Gregor, Sh., "Enterprise Architectures and Information Systems Alignment: policy, research and future implications" (2002). *ACIS 2002 Proceedings*. Paper 30. http://aisel.aisnet.org/acis2002/30

Martin, R. and E. Robertson (2002) Frameworks: Comparison and correspondence for three archetypes.

Martinette, C. V. and D. H. Dunford (2004). "Creating an empowered organisation." Fire engineering. Retrieved 7th April, 2011.

Martinsons, M. G., R. Davison, et al. (1998). "The balanced scorecard: A foundation for the strategic management of information systems." Decision Support Systems **25**.

Mathison, S. (1988). "Why triangulate?" Educational Researcher **17**(2): 5.

Maylor, 2005, Project Management, Third Edition with CD Microsoft Project, Prentice Hall, UK, p288

Maylor, H., T. Brady, et al. (2006). "From projectification to programmification." International Journal of Project Management **24**: 663-674.

McFarlan, F. W. (1984). The Information Systems Research Challenge, Harvard Business School Press.

McFarlan, F. W. (1984) The information systems research challenge.

McFarlan, F. W. (1995). Information-enabled organisation transformation oand outsourcing. Wirtschafts-informatik. P. D. W. Konig. Frankfurt, Instiut fur Wirtschaftsinformatik: 3-24.

McLean, E. and J. Soden (1977) Strategic planning for MIS.

McNamara, C. (2005). Field guide to consulting and organizational development with nonprofits. Minneapolis & Toronto, Authenticity Consulting, LLC.

Merkhofer, L. (2011) Project priority systems.

Middleton, D. P. (2004). "Alignment individual and organizational values to facilitate information systems adoption." Management of Computing and Information Systems WISER '04. Nov. 5, 2004: 72-77.

Miles, M. B. and A. M. Huberman (1984). Qualitative data analysis: A sourcebook of New methods. USA, SAGE Publications.

Miles, M. B. and A. M. Huberman (1994). Qualitative data analysis. London.

Mingay, S. (2007) Green IT: Dealing wit the shockwave.

Mintzberg, H. (1979). " An emerging strategy of "direct" research." Adminstrative Science Quarterly **24**: 582-589.

Mintzberg, H. (1993). Structure in fives: Designing effective organisations. NJ, US, Prentice-Hall, Inc.

Mirtidis, D. and V. Serafeimidis (1994). Evaluating information technology investments in Greece. 1st European Conference on IT Investment Evaluation, Henley.

Misuraca, G. (2010) IPTS exploratory research on emerging ICT-enabaled governance models in EU cities (EXPGOV). IPTS Information Society Unit

Mitchell, J. C. (1983). "Case and Situation Analysis." Sociological Review **3**(2): 187-211.

Mocnic, D. (2010) Achieving increased value of customers through mutual understanding between business and information system communities. Managing Global Transitions **8**, 17

Mocnic, D. (2011) Achieving increased value of customers through mutual understanding between business and information system communities. Managing Global Transitions **8**, 17

Moon, M. J. and S. Bretschneider (1997). "Can state government actions affect innovation and its diffusion? an extended communication model and empirical test." Technological Forecasting and Social Change **54**(1): 57-77.

Moore, M. H. (1995). Creating Public Value: Strategic Management in Government, Harvard University Press.

Morgan, B. C. and W. A. Schiemann (1999). "Measuring people and performance: Closing the gaps." Quality Progress **32**(1).

Morse, J. M., M. Barrett, et al. (2002). "Verification strategies for establishing reliability and validity in qualitative research." Information Journal of Qualitative Methods **1**(2): 9.

Motjolopane, I. and I. Brown (2004). Strategic business-IT alignment, and factors of influence: A case study in a public tertiary education institution. SAICSIT 2004, @ 2004 SAICSIT.

Muniz, E. J. S. (2009). "The investment in management information systems and its effect on the eficiency in public organisation when putting into practice the alignment fuction of infromation technologies governance." International Journal of Management and Information Systems **13**(2): 21-28.

Murray, R., J., and D. E. Trefts (2000). "Aligning IT and business strategy: The IT imperative in business transformation." Information Systems Management: 7.

Myers, M. D. (1997). "Qualitative research in information systems." MIS Quarterly **21**(2): 241-242.

Myers, M. D. and D. Avison (2002). Qualitative Research in Information System. London, SAGE.

Nachmias, C. F. and D. Nachmias (1996). Research methods in the social sciences. London: Arnold, St. Martin's Press, Inc.

NASCIO (2003). NASCIO Enterprise Architecture. Lexington, Kentucky, @ December 2003 NASCIO: 16.

NASCIO (2005). Government information sharing: Calls to action. U. S. D. o. Justice. USA, @ 2005 NASCIO. **1**: 63.

NASCIO (2006). Looking to the future: Challenges and opportunities for government IT project management officers. N. R. C. I. O. o. t. State. USA: 20.

NASCIO (2006) NASCIO survey on project management.

NASCIO (2007). Getting started in cross-bondary collaboration: What state CIOs need to know Representing Chief Information Officers of the States. Lexington, KY, USA, NASCIO: 10.

NASCIO (2007). IT Security Awareness and Training: Changing the Culture of State Government. R. C. I. O. o. t. States. Lexington, KY 40507, @ 2007 NASCIO.

NASCIO (2008). IT governance and business outcomes: A shared responsibility between IT and business leadership. IT Governance, The National Association of State Chief Information Officers.

Nelson, K. M. and J. G. Cooprider (1996). "The contribution of shared knowledge to IS group performance." Mis Quarterly **20**(4): 409-429.

NewsRx (2011) Energy Weekly News

Niederman, F., J. Brancheau, et al. (1991). "Information systems management issues fro the 1990s." Mis Quarterly **15**(4): 475-495.

Norman Vargas, Leonel Plazaola, et al. (2008). A consolidated strategic business and IT alignment representation: A framework aggregated from literature. Proceedings of the 41st Hawaii International Conference on System Sciences - 2008, Hawaii, Japan.

OGC (2008). OGC gateway review for programmes and projects. O. o. G. Commerce, www.ogc.gov.uk/what_is_ogc_gateway_review.asp.

Oh, W. and A. Pinsonneault (2007). "On the assessment of the strategic value of information technologies: Conceptual and analystical approach." Mis Quarterly **31**(2): 26.

Ojo, A., M. Shareef, et al. (2009) Strategic alignment: Aligning organisational and IT strategies. Centre for Electronic Governance

Palmer, J. W. and M. L. Markus (2000). "The performance impacts of quick response and strategic alignment in specialilty retailing." Information Systems Research **11**(3): 241-259.

Papp, R. (1995). Determinants of strategically aligned organisations: A multi-industry, multi-perspective analysis. Hoboken, NJ, Stevens Institute of Technology.

Pardo, T. A. (2009). Creating enhanced enterprise information technology goverfnance for New York State: a set of recommendations for value-generating. Center of Technology in Government. NY, University at Albany.

Pardo, T. A. and L. Dadayan (2006). Service New Brunswick. Public ROI - Advancing return on investment analysis for government IT: Case study series. NY, University at Albany, SUNY: 14.

Pare, G. (2004). "Investigating Information Systems with Positivist Case Study Research." Communication of the Association for Information Systems **13**: 233-264.

Patton, M. Q. (2002). Qualitative Research and Evaluation Methods. London, Sage Publication, Inc.

Pereira, C. M. and P. Sousa (2005). Enterprise architecture: business and IT alignment. 2005 ACM Symposium on Applied Computing, Santa Fe, New Mexico, ACM, New York, USA.

Perl, E. J. and D. F. Noldon (2000). Overview of student affairs research methods: Qualitative and quatitative. New Directions for Institutional Research, Wiley. **2000:** 37.

Pervan, G. (1998). "How chief executive officers in large organisations view the management of their information systems." Journal of Information Technology **13**: 95-109.

Peterson, R. R. (2001). "Information Governance: An empirical investigation into the differentiation and integration of strategic decision-making for IT." Tilburg University, The Netherland.

Phillips, J. (2003). PMP Project Management Professional Study Guide, McGraw-Hill Professional.

Pollalis, S. N. (1993). Computed-aided project management: A visual scheduling and management system, Wiesbaden.

Porter, M. E. and B. E. Millar (1985). "How information gives you competitive advantage." Harvard Business Review **63**(4): 149-160.

Powner, D. (2008), "Information technology OMB and agencies need to improve planning, management, and oversight of projects totaling billions of dollars", GAO-08-1051T, Washington, DC, July 31.

Presley, A. (2006). "ERP investment analysis using the strategic alignment model." Emerald Management Research News **29**(5): 273 - 284.

Queensland Government (2004). Charter of Social and Fiscal Responsibility.

Raghunathan, B. and T. S. Raghunathan (1990). "Planning implications of the information systems strategic grid: An empirical investigation." Decision Sciences **21**(2).

Rao, T. P. R., V. V. Rao, et al. (2004) E-Governance Assessment Frameworks: EAF version 2.0. E-Governance 49

Raven, R. P. J. M. (2008) The contribution of local experiments and negotiation processes to field-level learning in emerging (niche) technologies. Bulletin of Science, Technology and Society **28**, 13

Reich, B. H. and I. Benbasat (1996). "Measuring the linkage between business and information technology objectives." Mis

Quarterly **20**(1): 55-81.

Reich, B. H. and I. Benbasat (2000). "Factors that influence the social dimension of alignment between business and information technology objectives." MIS Quarterly **24**(1): 81-113.

Reichheld, F. (1996). The loyality effect: The hidden force behind growth, profits, and lasting value. Boston, Harvard Business School Press.

Renaud and Walsh (2010). 'The lost dimensions of strategic alignment'. Proceedings conference MCIS 2010 http://aisel.aisnet. org/mcis2010/70

Riege, A. and N. Lindsay (2006). "Knowledge management in the public sector: Stakeholder partnerships in the public policy development." Journal of Knowledge Management **10**(3).

Robinson, A. G. and S. Stern (1997). Corporate creativity: How innovation and improvement actually happens. San Francisco, Berrett-Koehler Publishers, Inc.

Rockart, J. F. (1979). "Chief executives define their own data needs." Harvard Business Review **57**(2).

Rockart., J. F. (1979). "Chief executives define their own data needs." Harvard Business Review **57**(2): 81-93.

Rolfe, G. (2006). "Validity, trustworthness and rigour: quality and the idea of qualitative research." Journal of Advanced Nursing **53**(3): 6.

Sabharwal, R. and Y. Chan (2001). "Alignment between business and IS strategies: a configurational approach." Information Systems Research **12**(1): 11-33.

Sabherwal, R. and Y. E. Chan (2001). "Alignment between business and IS strategies: A study of prospectors, analysers, and defenders." Information Systems Research **12**(1): 11-33.

Saul, F. and C. Zulu (1994). "Africa's survival plan for meeting the challenges of information technology in the 1990s and beyond." Libri **44**(1): 77-136.

Schniederjans, M. J. and J. L. Hamaker (2003). "A new strategy information technology investment model." Management Decision **41**(1): 5-17.

Schwalbe, k. (2006). Information technology project management, Thomson Course Technology.

Seddon, P. B., V. Graeser, et al. (2002). "Measuring Organizational IS Effectiveness: An Overview and Update of Senior Managemnt Perspectives." The DATEBASE for Advance in Information Systems **33**(2): 11-28.

Segers, A. H. and V. Grover (1999). "Profiles of strategic information systems planning." Information Systems Research **10**(3): 199-232.

Sherer, S. and S. Alter (2004). "Information system risks and risk factors: Are they mostly about information systems?" Communications of the Association of Information Systems **14**: 26-64.

Shimizu, T., M. M. de Carvalho, et al. (2005). Alignment of organizational strategy with information technololgy strategy. Strategic alignment process and decision support systems: Theory and case studies, IRM Press,2005.

Shin, N. (2002). "The impact of information technology on financial performance: the importance of strategic choice." European Journal of Information Systems **10**(4): 227-236.

Shpilberg, D., S. Berez, R. Puryear & S. Shah (2007). "Avoiding the Alignment Trap in Information Technology", *MIT Sloan Management Review*, Vol. 49, No. 1.

Siau, K. and Y. Long (2004). Factors impacting e-government development. International Conference on Information Systems

2004, ICIS 2004.

Sim, J.; Wright, C. C. (2005). "The Kappa Statistic in Reliability Studies: Use, Interpretation, and Sample Size Requirements". *Physical Therapy* **85** (3): 257–268.

Simons, G. F., K. L. A., et al. (2010). Enterprise architecture as language. Complex Systems Design and Management. M. Aiguier, F. Bretaudeau and D. Krob. Verlag Berlin Heidelberg, @ 2010 Springer. Proceedings of the 1st International Conference on Complex Systems Design and Management:18.

Singh, N., K.-h. Lai, et al. (2007). "Intra-Organizational Perspectives on IT-Enabled Supply Chains." Communication of the ACM **50**(1): 59-65.

Slaughter, S. A., L. Levine, et al. (2006). "Aligning software processes with strategy." Mis Quarterly **30**(4): 27.

Sledgianowski, D. and J. N. Luftman (2005). "IT-Business strategic alignment maturity." Journal of Cases on Information Technology **7**(2): 102-120.

Soeparman, S., H. v. Duivenboden, et al. (2009) Infomediaries and collaborative innovation: A case study on information and technology centered intermediation in the Dutch employment and social security sector. Information Polity 17

Stake, R., E. (2005). Qualitative case studies. CA, SAGE Publications.

Stake, R. E. (1995). The art of case study research, thousand oaks, SAGE publications.

Standing, C., A. Guilfoyle, et al. (2006). "The attribution of success and failure in IT projects." Emerald: Industrial Management and Data Systems **106**.

Stenbacka, C. (2001) Qualitative research requires quality concept of its own. Management Decision **39**, 6

Stevens, K. J. (2011). An investigation of risk management methodology use on information technology projects. School of Information Systems, Technology and Management, The University of New South Wales: 277.

Stewart, R. A. (2008). "A framework for the life cycle management of information technology projects: Project IT." International Jounal of Project Management(26): 203-212.

Strauss, A. and J. Corbin (1990). Basics of qualitative research: Techniques and procedures for developing grounded theory. Newbury Park, CA, Sage Publications, Inc.

Strauss, A. and J. Corbin (1998). Basic of qualitative research:Techniques and procedures for developing grounded theory. United Kingdon, SAGE Publications, Inc.

Strijbos, J.; Martens, R.; et al (2006) "Content analysis: What are they talking about?" *Computers & Education* **46**: 29–48.

Subramani, M. R., J. C. Henderson, et al. (1999) Linking IS-user partnerships to IS performance: A socio-cognitive perspective.

Suchan, J. (2003). "Define your project goals and success criteria." Retrieved 7th April 2011, 2011, from http://office.microsoft.com/en-us/project-help/define-your-project-goals-and-success-criteria.

Sullivan, C. H. (1985). "Systems planning in the information age." Sloan Managaement Review **26**(2): 3-12.

Tallon (2000), Understanding the dynamic of information management costs, Communications of the ACM vol. 53 (5), p. 121-125

Tallon, P. P. and K. L. Kraemer (2003). Investigation the relationship between strategic alignment and IT business value: The discovery of a paradox. Creating business value with information technology: Challenges and solutions. N. Shin, Idea Group Publishing: 22.

Tan, C. W. and S. L. Pan (2003). "Managing e-transformation in the public sector: An e-government study of the Inland Revenue Authority of Singapore (IRAS)." European Journal of Information Systems 12: 12.

Tan, F. B. and R. B. Gallupe (2006). "Aligning business and information systems thinking: A cognitive approach." IEEE Transaction on Enginnering Management 53(2): 223-237.

Tang, A., J. Han, et al. (2004). A comparative analysis of architecture frameworks. Technical Report SUTIT-TR2004.01. Swinburne, Swinburne University of Technology.

Taylor-cummings, A. (1998). "Bridging the user-IS gap: A study of major information systems projects." Journal of Information Technology 13: 29-54.

Templeton, G. F., B. R. Lewis, et al. (2002). "Development of a measure for the organizational learning construct." Journal of Management Information Systems 19(2): 175-218.

The World Bank (2007). Public value of IT frameworks.

Thomas, J. and M. E. Mullaly (2007). "Understanding the value of project management: First steps on an international investigation in search of value." Project Management Journal 38(3): 74-89.

Tremblay, M. (2006). "Knowledge as power in the public sector." Content Management. Retrieved 7th April 2011, 2011.

Tzeng, S.-F., W.-H. Chen, et al. (2008). "Evaluating the business value of RFID: Evidence from five case studies." International Jounal of Production Economics 112: 12.

Valorinta, M. (2011). "IT alignment and the boundaries of the IT function." Journal of Information Technology 26(1): 46-59.

van Eck, P., H. Blanken, et al. (2004). "Project GRAAL: Towards operational architecture alignment." International Jounal of Cooperative Information Systems 13(3): 235-255.

Verner, J. and W. M. Evanco (2005). "In-House Software Development: What Project Management Practices Lead to Success?" IEEE Software 22(1): 86-93.

Vogt, M. and K. R. Hales (2010). Strategic Alignment of ICT Projects with Community Values in Local Government. Proceedings of the 43th Hawaii International Conference on System Sciences, Hawaii Island, IEEE Computer Society.

Walsh, I. and A. Renaud (2010). "La theorie de la traduction revistee ou la conduite du changement traduit." Management et Avenir, Special Issue (a paraitre).

Wang, W.-T. and S. Belardo (2009). "The role of knowledge management in achieving effective crisis management: A case study." Journal of Information Science 35(6): 24.

Ward, J. and J. Peppard (2002). Strategic planning for information systems. Bedfordshire, UK, John Willey & Sons, Ltd.

Warland, C. (2005). Awareness of IT control frameworks in an Australian State Government: A qualitative case study. the 38th Hawaii International Conference on System Sciences, IEEE.

Watson, R. T., G. G. Kelly, et al. (1997). "Key issues in information systems management: An international perspective." Journal of Management Information Systems 13: 91-115.

Weerakkody, V., J. Janssen, et al. (2007). "Integration and enterprise architecture challenges in E-government: A European perspective." International Journal of Cases on Electronic Commerce 2(3): 13-35.

Weilbach, L. and E. Byrne (2009). Aligning national policy imperatives with internal information systems innovations: A case study of an open source enterprise content management systemin the South Africa public sector. The 10th International Coference on Social Implications of Computers in Developing Countries, Dubai, United Arab Emirates, e-publications@RCSI.

Weill, P. and J. W. Ross (2004). IT governance. USA, Harvard Business School Publishing, 60 Harvard Way, Boston, Massachusetts 02163.

Whittaker, B. (1999). "What went wrong> Unsuccessful information technology projects. ." Informat Manag Comput Security **7**(1): 23-29.

Whitten, J. L., L. D. Bentley, et al. (2004). Fundamentals of Systems Analysis and Design Methods, The McGraw-Hill companies, Inc.

Wieringa, R. J., P. A. T. van Eck, et al. (2004). Architecture alignment in a large government organisation: A case study. AE enschede, the Nertherlands, Center for Telematics and Information Technology.

WIKIPEDIA (2011). Project management. WIKIPEDIA, Wikimedia Foundation, Inc.

Wilkin, C. L. and J. Riddett (2009). "IT governance challenges in a large not-for-profit healthcare organisation: The role of interanets." Electron Commer Res **9**.

Williams, P. A. (2005) IT alignment: Who is in charge?

Winter, G. (2000). A comparative discussion of the notion of 'validity' in qualitative and quantitative research. The Qualitative Report. **4**: 58 paragraphs.

Yaghoubi, N.-M. (2010). "Identifying IT-based rural development infrastructures: Strategic alignment approach." Journal of US-China Public Administration **7**(5): 6.

Yang, J., G. Q. Shen, et al. (2010). "Stakeholder management in construction: An impirical study to address research gaps in previous studies." International Project Managemetn Association **7**(13).

Yin, R. K. (1988). Case study research: Design and methods. United States of America, SAGE Publications, Inc.

Yin, R. K. (2003). Applications of case study research, Beverly Hills, CA, SAGE Publications.

Ylimaki, T. and V. Halttunen (2005/06, p.189). "Method engineering in practice: A case of applying the Zachman framework in the context of small enterprise architecture oriented projects." Information Knowledge Systems Management **5**: 20.

Yow, V. R. (1994). Recording oral history: A practical guide for social scientists. United Kingdom, SAGE Publications, Inc.

Zarvic, N. and R. Wieringa (2004). "An integrated enterprise architecture framework for business-IT alignment." 2007.

APPENDICES

APPENDIX A: GATEKEEPER LETTER

Abdullah Al-Hatmi
Faculty of Business, Technology and Sustainable Development
Bond University,
QLD 4229
Australia
Tel: +61- (0)430963330

Subject: **BUHREC Protocol Number RO-1016**

'Council' is pleased to accept your request to be a research partner in the ICT governance processes, structures and relational mechanisms, and their impact on IT projects. The research project is undertaken at the School of Information Technology, Bond University.

We understand that the project will involve interviews with key IT members holding pivotal roles in the ICT governance context. The interviews will be conducted at a location agreed to by mutual consent and will last for one to two hours.

We understand that the following conditions will be adhered to:

- Business/participants' confidentiality will be maintained throughout the research. Participant identity will only be known to the researcher and will not be revealed to the **'Council'** or its representatives.

- Participation is voluntary, and participants are free to withdraw at any stage without providing any reason. Withdrawal from the study will not be penalised in any way.

While the data collected will be used for research purposes and transcripts will be verified by participants and organisation, information provided by the researchers to the council will be at the group level only or de-identified.

At **'ICT Governance in Council',** we value and uphold the pursuit and advancement of knowledge through research and education, especially research that results in providing practical value to practitioners in real life settings. We trust that as ICT governance practitioners, our participation will provide valuable insights into a greater understanding of the perpetuation of IT project success across generations in public organisations.

Authorised by: _____
(Please print name)
Signature: _____
Company Position: _____
Date: _____

APPENDIX B: APPROVAL LETTER

An Analysis of ICT Strategic Alignment in Public Agency

Executive Summary

This ground-breaking research aimed at the public agency, offer insights from the IT governance

HUMAN RESEARCH
ETHICS COMMITTEE
Bond University
Gold Coast, Queensland 4229
Australia

Ph: +61 7 5595 4194
Fax: +61 7 5595 120
(from overseas)

Email: buhrec@bond.edu.au

ABN 88 010 694 121
CRICOS CODE 00017B

8 October 2009

Asst Prof Kieth Hales/Abdullah Al-Hatmi
Faculty of Business, Technology and Sustainable Development
Bond University

Dear Kieth and Abdullah

Protocol No: **RO1016**
Project Title: **An Analysis of ICT Strategic Alignment in Public Agency**

I am pleased to confirm that you have now satisfied the requirements of BUHREC after an Expedited Review of your project. You may commence your research.

It is important to remember that BUHREC's role is to monitor research projects until completion. The Committee requires, as a condition of approval, that all investigations be carried out in accordance with the National Health and Medical Research Council's (NHMRC) National Statement on Ethical Conduct in Research Involving Humans and Supplementary Notes. Specifically, approval is dependent upon your compliance, as the researcher, with the requirements set out in the National Statement.

Additionally, approval is given subject to the protocol of the study being under taken as declared in your application, with amendments, where appropriate.

As you may be aware the Ethics Committee is required to annually report on the progress of research it has approved. We would greatly appreciate notification of the completed data collection process and the study completion date.

Should you have any queries or experience any problems, please liaise directly with Caroline Carstens early in your research project: Telephone: (07) 559 54194, Facsimile: (07) 559 51120, Email: buhrec@bond.edu.au.

We wish you well with your research project.

Yours sincerely

Dr Mark Bahr

APPENDIX C: CONSENT FORM

Consent Form

Re: An Analysis of ICT Strategic Alignment in Public Agency: A Case Study in Council

I_____ agree to participate in the research project entitled 'An Analysis of ICT Strategic Alignment in Public Agency: A Case Study in Council', being conducted by Abdullah Al-Hatmi, School of Information Technology, Bond University.

I understand that the purpose of this study is to understand Strategic Alignment (SA) perspectives required for better achievement of value realisation in public organisations.

I understand that my participation in this research will involve undertaking one or two in-depth interviews lasting approximately 1 hour at a venue to be agreed upon between Abdullah Al-Hatmi and me. The interview process may be supplemented with telephone conversations that will take place at my convenience between Abdullah Al-Hatmi and me.

I understand that I may choose to discontinue my participation at any time for any reason.

I give permission for Abdullah Al-Hatmi to use a tape recorder to record the interview and make brief additional notes during the interview to clarify understanding and to assist with his later analyses and interpretation.

I agree that the research data gathered from this project may be published in a form that does not identify me in any way.

Name _____
Position_____
Work Phone*_____
Signature of Participant_____
Date_____

* This number will be used in strictest discretion. No mention of this study or your experiences will be made to anyone who may answer your telephone.

NOTE: This study has been approved by Bond University Human Ethics Review Committee. If you have any complaints or reservations about the ethical conduct of this research, you may contact the Ethics Committee through the Research Ethics Coordinator (+61 7 559 54 194). Any issues you raise will be treated in confidence and investigated fully, and you will be informed of the outcome.

www.bond.edu.au

APPENDIX D: EXPLANATORY STATEMENT FOR RESEARCH PARTICIPANTS

Explanatory Statement for Research Participants

RO-1016: An Analysis of ICT Strategic Alignment in Public Agency: A Case Study in Council

Dear Participants,

This explanatory statement is written to inform you about the conditions of your participation in a research project to be undertaken at Council.

1. RESEARCH FOCUS: While public agencies deliver their IT projects on time and on budget, few report that they actually identify value realisation gained from IT projects. Scant attention has been paid to the Strategic Alignment (SA) perspectives and IT governance to improve the return of government IT projects. Hence, it is the focus of the research.

2. **CONTACT DETAILS**: I can be reached by e-mail at aa418@hotmail.com.

3. **PARTICIPANTS**: Council ICT Governance staff will be invited to participate in the research study. The first aspect of the study requires participants to read this Explanatory Statement, which describes the research process. It should take no longer than five minutes. The completion of this interview should take no longer than 60 minutes.

4. **VOLUNTARY PARTICIPATION**: Participation in this research is entirely voluntary. If you agree to participate, you have the option to discontinue your involvement at any time prior to submitting your responses.

5. **INFORMED CONSENT**: After reading this Explanatory Statement, you will be invited to sign a Consent Form to indicate your willingness to participate in this research.

6. **ANONYMITY**: Please be assured that your responses will be kept strictly confidential (i.e. collected in a de-identified form). Individual participants will not be identified in the analysis, as only thematic/categorised results will be analysed and presented in academic journal publications and/or conference proceedings.

7. **COLLECTED DATA**: Bond University has clear guidelines on the procedures concerning the storage of research data. The information you provide for this research will be kept in a locked filling cabinet for five years from the completion of the research, after which it will be destroyed.

8. **BOND UNIVERSITY COMPLAINTS CLAUSE**: Should you have any complaints concerning the manner in which this research (RO-1016) is conducted, please do not hesitate to contact Bond University Human Research Ethics Committee (BUHREC) at the following address:

Mrs. Caroline Carstens
HDR Administrator and Ethics Officer
Bond University Research and Consultancy Services
Bond University QLD 4229
Australia
Tel: +61 7 559 54194 Fax: +61 7 559 51120
Email: ccarsten@bond.edu.au

9. **ACCESS TO RESULTS**: If you are interested in finding the overall results of this research, you can do so by e-mailing kihales@bond.edu.au or abalhatm@bond.edu.au

Thank you for your willingness to participate in this research.
With kind regards and best wishes,

Prof. Kieth Hales
Assistant Professor
School of Information Technology
Bond University
+61 7 559 53356
kihales@bond.edu.au

Prof. Iain Morrison
Head of School of Information Technology
Bond University
+61 7 559 53359

imorriso@bond.edu.au

Appendix E: Research background

An Analysis of ICT Strategic Alignment in Public Agency

Executive Summary

This ground-breaking research aimed at the public agency, offer insights from the IT governance perspectives towards public value commitment of IT projects. The intended respondents are at senior managerial level, with the survey results reflecting how Council enable both business and people to executive their responsibilities in support of business/IT alignment and the creation of business value from corporation investments.

* Effective IT Governance is the single most important predictor of the value on organization generates from IT'   (Weill & Ross, 2004).

* For the first time we'll be looking at a whole of government framework to guide the development of our information systems...the framework is going to ensure the decisions about ICT right across the government align with government priorities'.

The Minister of Queensland IT
(Renai LeMay, ZDNet Australia, 2006)

* SA is a high priority topic in executives' agenda in Europe and America
(CSC,2000; PriceWaterhouse, 1996)

* Strategic alignment proves to be one of the most important issues facing future organizations.   (Tallon & Kraemer 2003)

Future Trends

Though technologies have potential to improve the lives of people in the world, however, 'the best practice frameworks of IT Governance in governments have not been fully investigated' (Yining Chen, Wayne W. et al 2007)

- What decisions must be made?
- Who should make these decisions?
- How will we make and monitor these decisions?
- How to prioritize those decisions?

Indicators of the Increased IT Governance Trend

- More regulations and legislative frameworks response to the continued high rate of IT project failure such as COBIT, Val IT, ITIL, AS 8015-2005, ISO 20000, PMBOK, PRINCE2 and COSO
- Increase in partnerships, accountability and clarity of direction amongst top IT executives.
- More public frameworks emerge to address non financial return of IT investments in governments such as MAREVA in France, Gartner' BVIT, WiBI in German, Federal EA Performance Reference Model, DAM & VAM in Australia.

176

Project Plan

Research Objective

- to investigate the role of SA in the successful of IT projects

- How is Strategic Alignment (SA) achieved in a public organization?

Research Approach

- Case study in Council
- Semi-structured interviews (60 mins per session per person)
- Analysis of public documents, reports and minutes meetings.
- Literature Review (both academic & non-academic)
- In-depth interviewing

Ways Council Can Gain Advantage

Trends indicate that government IT projects success remains critical in the future. Council's participation in those conversations will be crucial to create the most effective public policies possible.

Public agencies that sustain strategic alignment over time secure better IT governance practices. Sharing decision-making and accountability will increase clarity and direction of Council strategy which in turn will enhance an overall governance performance.

Prepared by: Dr. Abdullah Al-Hatmi

Council Commitment

- 24 hours of interview time
  (interview with key personnel at least middle or senior managerial level for 60 minutes each)
  (participants to be identified by both researcher and Council)
- Low level of supervision as part of research approach (audio recorded & transcribed)
- Proposed interview period (Sept., 2009)
- Interview sessions @ Council's office during data collection period (subject to availability)
- Project Deliverables (March, 10)
  a corporate-styled findings report will be presented to Council

Required Data

How IT projects in Council is evolving through the use of improve governance process.

Specific information needs:
* Proposals plan,
* business cases,
* post implementation review,
* ICT Portfolio Management, and
* Corporate Activity.

There is need for practical hands-on info:
- How IT projects and public value are being monitored and measured

APPENDIX F: LIST OF GOVERNMENT DOCUMENTS AND REPORTS

The following documents and reports were provided during the course of the study:

- Business Case
- Project Management Plan
- Project Charter
- Value Realisation Plan
- Post-Implementation Review (also known as Learning Report)
- Value Assessment (sometimes called Value Profile)
- Status Report
- Quarterly Report Issues
- Quarterly Benefits Report
- Project Handover Report
- Project Closure Report
- Consolidated Report
- Agenda
- CGC Minutes Meeting
- Enterprise Architecture
- Information Management Policy
- Policy Framework Summary Review
- Information, Communication and Technology (ICT) Policy Framework and Policy Structure Overview
- Project Management Policy and Procedure
- Internal Audit Report
- Quality Management Plan
- Detailed Mapping C OBIT to ICT Policy Framework
- Research Report
- ICT Strategy - 2005 to 2009

APPENDIX G: INTERVIEWS PROTOCOL

Interviewee:		Date:	
Questions Area Addressed		**Points to cover**	**Notes**
0	Warm up	▪ Set up Ethics form ▪ Tape recorder	
1	Participant & Organization background	▪ Job title and experience ▪ Historical ICT Strategy development ▪ IT functions in organization	
2	IT Governance context	▪ Roles of leadership ▪ The importance of IT governance in business/IT function ▪ The vision of IT governance and its principles	
3	The Concept of SA in general	▪ Definition of SA and its importance to organization ▪ SA aspects	
4	IT projects	▪ IT project management ▪ The definition of project success ▪ CSFs of IT projects ▪ Management issues	
5	**Strategic Alignment Perspectives** ▪ Strategy: ▪ Knowledge ▪ Decision-Making ▪ Enterprise Architecture ▪ Public Value	▪ IT/business strategy, alignment, goals and objectives ▪ Performance of IS ▪ Business/IT communication ▪ Reporting mechanisms ▪ How decision is made in IT functions ▪ Outsourcing decision ▪ Components of IT architecture ▪ Business/IT architecture solutions ▪ ROI and financial metrics ▪ Tools to measure BVIT	
6	Closure	▪ Advice to improve the success rate of IT projects ▪ Thank participant & turn off recorder ▪ Request for the possibility of follow up on questions	

APPENDIX H: INTERVIEW QUESTIONS

Participant and Organisation Background

- What would you consider some of the key changes in ICT Strategy over the last 5 years that have had the most impact on Council?

- How are Council organized in terms of the IT function? Who does the top IT executive or CIO report to?

- What would you consider the most important factor in improving the ability of IT to play a larger role in Council?

- What is the role of organizational politics and power in enabling or inhibiting the strategic processes of an IT function?

IT Governance context

- Describe the principles of IT governance and its importance to a Council.
- What are the benefits and drawbacks of a matrix organization?
- Why is, how IT is organized, important to the entire organization?
- Why is, how the business is organized, important to the IT organization?
- Can IT be organized centrally while the business is organized de-centrally? And vice-versa?
- Have you observed any symptoms of poor IT governance in Council? What are they?
- Describe some alternative forms of governance and their advantages and drawbacks.
- Why is governance important to effective management of cross-functional initiatives?
- Describe the primary leadership roles and responsibilities in IT governance.
- What considerations regarding IT governance come into play when addressing ICT strategic directions?

The Concept of SA in general

1. Why is the state of alignment maturity important to an organization?

The Concept of SA in general

IT projects

- How does your organization evaluate project risk?
- How does your organization de-escalate troubled projects?
- What techniques might an IT steering committee use to prioritize projects?
- Amongst 14 projects, I am working with, milestones and deliverables have not been reached as expected due to one reason or another. How does Council maximize the success of delivery IT projects in the future?
- What are some of the Emerging Technologies of 5 years ago? now? Which were successful, which unsuccessful? Provide reasons

- What mechanism does your IT organization have in place regarding vendors? How well do they work?

- How are outsourcing decisions made in your organization? By whom?

- What governance mechanisms are most commonly employed in managing vendor relationships?

- Describe the decision-making process when considering outsourcing arrangements.

- What are the primary roles of an IT steering committee?

- Who should be part of an IT steering committee and why?

- In regard to IT project success, what does the term 'success' mean to you?

- Project 10 has suffered due to limitations such as inadequate project plan, knowledge gaps of project members and the delay of Steering Committee decisions which were then not actioned. The cost of delayed decision was estimated to be approximately $20,000 per week.

- How do you see the differences in the process mechanisms in the past and now to avoid such problem in the future?

Strategy

- Does Council have an IT strategy? What are its strategic goals and objectives?

- Name some symptoms of lack of alignment between IT and the business.

- What tools do you use to measure the effectiveness of an IT strategy?

- How does a Council use IT to drive its business strategy?

- What tools do you use to measure the effectiveness of an IT strategy?

- How often has your IT management formally assessed its Management Systems Plan?

- What forms of security planning, business continuity planning and privacy planning is Council *mandated* by law to perform and test?

- Some say that the progress towards building a comprehensive framework for IT assessment has begun, but may never finish. Explain.

- Why is planning important for IT processes?

- What do you think are the main Council weaknesses in evaluating the performance of Information Technology?

- Why have disaster recovery processes and security processes gained prominence in our economy?

- Describe how the implementation of IT involves change management issues.

- What factors motivate change?

Knowledge

- What can a Council do to improve the education levels of their staff? What is the role of IT or business process certifications? Should continuing education and cross training be required?

- What are the negative implications and issues when firms attempt to cross train their business and IT staffs.

- What tools should be developed in the future to support improved business – IT communication?

- How might a CIO get more involved in the strategic processes of the business functions in Council?

- What is the importance of having good communications between IT and the business?

- What are the key elements for effective communications of vision?

- How does cross training or cross education improve communication?

- What are the benefits to the Council, if the business people understood IT better?

- In what ways can the Council improve business and IT communications? Explain.

- What are the key *management* skills and responsibilities of the CIO?

- What are reporting mechanisms deployed in Council to ensure the effectiveness of the life-cycle of the IT projects?

- Why business/IT planning is important?

- Communicating honestly with people is one part of establishing and maintaining trust and building a strong and valued reputation. How do the CIO, the IT staff, and their business counterparts in Council communicate effectively with each other, so that the firm can improve their business systems and reduce errors and waste.

Decision-Making

- How will your organization make decisions about its international IT functions?

- In project no 2 OCR Accounts Payable Project, the vendor was unable to deploy the software. The vendor inadequately interpreted the business requirement and underestimate configuration. What mechanism does your IT organization in Council have in place regarding vendors? How well do they work?

- How are outsourcing decisions made in your organization? By whom?

Enterprise Architecture

- How does Council align its strategic priority with business/IT architecture solutions?

- What are the components of IT Architecture in Council?

- How has productivity increased in application development?

- How can Operational layer processes be performed with much less staff than comparable tactical layer processes?

- Are there processes that span more than one layer?

Public Value

- What contributes to the difficulties IT organizations have in demonstrating the value technology brings to the business goals and results? What is the role of the business function in these difficulties?

- What are the major components of measurement that can be used to measure performance of Information Technology?

- Why are non quantifiable measures of performance better indicators than quantifiable measures? What intangible items would be evidence of effective performance indicators?

- Return on investment (ROI), has always been a main focus of financial analysts. What intangible benefits of Information Technology are often difficult to quantify for this measurement?

- Why are Return on investment (ROI) financial measures not often good indicators of performance?

- How does managing a portfolio of IT investments compare to looking at IT projects individually.

- Explain the concept of the Value Realization in Council.

- What are the differences between Activity-Based Costing and Activity-Based management?

- How does the Council adopt multiple value methodologies to manage their Information Technology function? Explain.

- Why isn't business and financial measures sufficient to value the effectiveness and level of contribution that Information Technology provides?

- Does your company have an effective set of tools to measure the value of IT investments?

APPENDIX I: MATURITY OF SA ATTRIBUTES IN PROJECTS

Tables 5-1 to 5-14 show the maturity levels of attributes. Descriptions of each attribute are given in the right-hand column. The maturity levels range from Ad Hoc (1) to Optimised (5).

TABLE 5-1: MATURITY OF SA ATTRIBUTES IN PROJECT 1

SAP	SA Attributes	Maturity					Rationale
		1	2	3	4	5	
Strategy	Clarity of direction		X				Clearly articulated and includes storing and designing water database.
Strategy	Performance measures		X				Measurement tools were inadequate.
Strategy	Quality of IT/business Plan		X				Business plan included insufficient detail.
Knowledge	IT/business manager communication		X				Communication was poor. Project manager was marked as 'TBA' in Steering Committed and Project Management team.
Knowledge	Organisation emphasis on knowledge		X				Report tools were used via the reporting/responsibility hierarchy to spread knowledge, but the tools were used immaturely.
Knowledge	Training		X				Staff training was not carried out.
Decision Making	IT investment and budgeting				X		IT investments and alternative costs were taken by the CGC, the Steering Committee and the Project Management Team. With the exception of assigning strategic and budget decisions to the Corporate Governance Committee, the project's hierarchy did not indicate by whom and how other decisions are taken.
Decision Making	Prioritisation			X			High priority was given to the ICT Strategy and Corporate Plan; its implementation was deemed necessary to the Council and to the society; business-driven.
Decision Making	Stakeholders			X			Twenty stakeholders were identified from different business units from the Council. Eight were key stakeholders and only one, Business Information System, was engaged in both the Steering Committee and the Project Management group. There were not stakeholders from outside or community representatives involved in any part of the process.

	SA Attributes	1	2	3	4	5	Rationale
Enterprise Architecture	**Aligned technical/ business solutions**		X				Project was not capable of integrating with other systems such as Oracle or Citipac Financial systems.
	Application and technology				X		Storage capacity, applications support and development resources were considered prior the implementation of phase 2. Ability to design and create database storage, Business Objects Universes and develop Business Objects reports.
	Risk assessment			X			A few risks were identified but were not deemed critical to the organisation: the absence of a stakeholder on the scheduled task and the fact that a test environment for the Business Objects application was not carried out earlier. All other risks were identified, including hardware capacity and lack of Applications Support Resources.
Public Value	**Benefits to organisation**			X			Report capability and decision in regards to Water Meter Renewals Management and Water supplied per property were planned. Reduced variance in queries from staff, gains in efficiency, insurance risk mitigation to the organisation and effectiveness in operational level were also considered in plan.
	Benefits to public		X				Extracted data contained accurate property and consumption data, water and billing applications with historical data over the past eighteen months. Through use of a unique key such as a property number, additional reports could be generated, providing information relating to the water consumption of properties where particular licensed operators occur.
	Economic/ financial metrics		X				Financial metrics were not strongly considered.

TABLE 5-2: MATURITY OF SA ATTRIBUTES IN PROJECT 2

SAP	SA Attributes	Maturity					Rationale
		1	2	3	4	5	
Strategy	**Clarity of direction**				X		Direction is clearly communicated: payment is accurate and on time. The objective was identified: to scan all invoices and post into Ellipse.
	Performance measures			X			Success criteria were not identified. Acceptance tools and phases were used and baseline plan were identified. Project manager formally addressed variation.
	Quality of IT/business plan			X			A good business/IT plan, including legislative requirements and project process, was scheduled to be delivered upon phases. PMP was created while the project was already in progress.

Knowledge	IT/business manager communication					The Steering Committee consisted of five members and featured business/IT communication. Participation was aligned with core competencies of business/IT strategy. There was no clear formal meeting between IT and business managers.
	Organisation emphasis on knowledge					Project matters were communicated to project manager via different forms, such as email, Microsoft Word, Microsoft Excel, Visio, PDF and meeting minutes. Issues, risks and actions were tracked and documented. Communication plan existed.
	Training					Training the users and the support personnel was mentioned as a strategic goal, but no training was given in practice.
Decision Making	IT investment and budgeting					Budget expenditure, budget approval, quotes, all purchases, Purchase Requisition Number, project budget holders and cost estimates were all addressed by the team and project manager. Ten per cent of the budget increased.
	Prioritisation					Key decisions about organisational priorities were taken by the PMP, the CGC and the Steering Committee. Vendors were selected without approval from the CGC. The project charter was used by senior/executives management to make decisions, and other decisions were taken by the group.
	Stakeholders					The project had a wide range of stakeholders, including suppliers, customers, project manager, business analyst, internal audit, senior accounts payable officers, record services, infrastructure support, business solutions support specialists, sign-off technical handover and many consultants. Vendors failed to deliver on time and a legal officer resigned from the Council.

		1	2	3	4	5	Rationale
Enterprise Architecture	Aligned technical/business solutions						Core competencies integrated between business and IT. The quality of customer service in IT and business was linked via internal and external data accuracy. Business solutions were integrated with Oracle and SQL.
	Application and technology						OCR Configuration was fully deployed to ensure the application worked. Infrastructure services, such as network, bandwidth and servers, were addressed. Unified recovery solutions, desktop services, hardware specifications, SAL and security level were defined to ensure appropriate information and services were obtained.
	Risk assessment						The project management group deployed a PMO Risk Management Plan Template and Risk Register to track risks. Individual risks were allocated a probability and impact assessment. Scope, budget and schedule were labelled critical, high-risk, medium-risk and low-risk, based on a set of criteria. Software did not work until late in the project. Internal audit, training testing and SAL were all delayed. Other out-of-scope functions were delivered, such as faxing and scanning.
Public Value	Benefits to organisation						Data-entry accuracy, operational effectiveness and efficiency gains were planned.
	Benefits to public						The goal of this project is to improve customer service and mitigate risk across the Council.
	Economic/financial metrics						Few financial metrics were provided.

TABLE 5-3: MATURITY OF SA ATTRIBUTES IN PROJECT 3

SAP	SA ATTRIBUTES	MATURITY					RATIONALE
		1	2	3	4	5	
Strategy	Clarity of direction						Objectives were clear: to implement an integrated user-friendly system and to provide a centralised system in which all relevant application details are captured and referenced against the records systems and the property database.
	Performance measures						Many key performance indicators were identified: accurate and reliable data, availability of reports, equitable distribution of work, staff take-up of the use of AMS and acceptable levels of alerts.
	Quality of IT/business plan						BC was poorly documented and lack of details (e.g. functional specification and business requirements) were broadly identified. Lack of PIR and estimated cost analysis was not available before the completion of Phase 1.

Knowledge	**IT/business manager communication**						Communication occurred in the Steering Committee and via the CGC.
	Organisation emphasis on knowledge						Responsibility sharing, resources plans, and formal communication plans were used to facilitate the implementation. The tender process, The Integrated Planning Act, analysis of functional specification, Request for Change Process and linked documents were used to enhance the skill knowledge brought to bear on the project.
	Training						The quality of training was reasonable: training plan, start and completion of training, users accepting tests, etc. The project needs improvement in creating standard mechanisms for post-training and user acceptance, supervision attendances and booking in advance.
Decision Making	**IT investment and budgeting**						Carried out by committees. Target date for completion was not identified, as it depended on Phase 1; financial analysis was not available for the same reason. Budgeting was carried out by individuals such as the Change Manager in RFG. Business owner approval and other key decisions were taken by committees such as the Change Advisory Board. Uncertainty of decisions occurred; the case of evaluation of an impact assessment, uncertainty conditions.
	Prioritisation						The project was selected with the Council's future integration needs and the necessity of centralised information in mind. Enhancements were prioritised and scheduled in Phase 1.
	Stakeholders						Stakeholder analysis identified responsibilities of business systems owners, steering committee, community service staff, customer service staff, end user group and trainers.

	SA Attributes	1	2	3	4	5	Rationale
Enterprise Architecture	Alignment of technical and business solutions			X			Application of lodging, tracking, process and governance were addressed to facilitate lodging applications via e-mail and copying information using Windows functionality. This in turn aided regular tracking of the status of any application. Rollout and technical problems called for deployment of document-linking and TRACKS.
	Application and technology		X				Low ability of integrated applications for town application. AMS document-linking did not cover all the requirements of the business, such as functionality of the standard building relaxation letters, GIS update interface, free calculation model and integration with Outlook and AMS/Citipac.
	Risk assessment		X				Risks were identified and description, likelihood, consequences, mitigation and scope of mapping integration were identified before analysis was conducted. Other risks identified include availability of staff, their resistance, delays in approval of funding or sign-off and customer satisfaction in free calculation.
Public Value	Benefits to organisation		X				Benefits were identified: a skeletal map of the IDAS system, establishment of a 'one-stop shop', accurate and reliable data. No VRP with measurement benefits was realised.
	Benefits to Public			X			Benefits included availability of suitable technology offered to customer services; accurate quotes for and receipt of customer fees; and. improved planning and management decisions.
	Economic/financial metrics	X					Lack of business/IT financial metrics.

TABLE 5-4: MATURITY OF SA ATTRIBUTES IN PROJECT 4

SAP	SA Attributes	Maturity					Rationale
		1	2	3	4	5	
Strategy	Clarity of direction						Objectives were defined: creating a business continuity plan (BCP); addressing the organisation's ability to continue functioning when normal operations are disrupted; and building disaster recovery, end-user recovery, contingency, emergency response, crisis management, as well as consistent and reliable customer service.
	Performance measures						Risks and cost estimation were provided, but milestones and deliverables were not.
	Quality of IT/business plan						The project has rich BC and PMP content titles. However, details about deliverables, cost/benefits and stakeholders management were not given. Details about BRP, revision history, distribution to key stakeholders and external resources were not specified.

Knowledge	IT/business manager communication						While stakeholders were identified in general, project structure lacked detail, and little is known about communication between business and IT managers.
	Organisation emphasis on knowledge						The need for change was given, and links with other projects/programs were explained. Roles/responsibilities were also addressed amongst stakeholders and impacted directorates.
	Training						Training was provided to the project's decision-making members. Initial training of relevant staff was required to activate the Corporate BCPs. Simulation and communication exercises training were put in place.
Decision Making	IT investment and budgeting						Resource estimates were carried out by a directorate representative, in accordance with a value profile and with Business Continuity Plans 2005-2006. Budget/funding and a value profile were taken into consideration when determining project expedition.
	Prioritisation						The process followed Corporate Strategic Priorities and Corporate Plan. Business continuity was deemed necessary in the event of a service interruption.
	Stakeholders						Decisions were communicated to key stakeholders.
	Systematic decision						Key stakeholders and impacted directorates were involved formally in decisions.
Enterprise Architecture	Aligned technical/business solutions						Integrated solutions pertaining to the level of risk associated with information security and Council's mitigation risk were dealt with cost effectively.
	Application and technology						Technical requirements were addressed, though some of the IT infrastructure and IT Disaster Recovery Plans required were out of the scope of project implementation. Network enterprise application included the Corporate Risk Register, various specific risk registers, the Risk Policy and Guidelines, the Business Continuity Plan and a Disaster Recovery Plan ensuring the various parts work in concert with each other and result in a consistent approach to the identification and mitigation of risk across the organisation.
	Risk assessment						Risks were identified in 'Corporate Risks and Issues', but the only plan for a rapid resumption of Council services offered solutions to mitigate a variety of risks, such as food shortage, bushfire, earthquakes, terrorism, disease, or tsunamis. Two risks were described in a table, and another table offered cost estimation. Other measurements, such as project deliverables, constraints and dependencies, did not take place.

Public Value	Benefits to organisation				▨		Service continuity was addressed.
	Benefits to public			▨			Benefits included endorsing critical operations and accountable persons; mitigating risks in the areas of social, environmental and economic sustainability; and improving public service in the event of service interruption and rapid resumption of these services.
	Economic/financial metrics		▨				Traditional cost estimation was used, without financial metrics.

TABLE 5-5: MATURITY OF SA ATTRIBUTES IN PROJECT 5

SAP	SA Attributes	Maturity					Rationale
		1	2	3	4	5	
Strategy	Clarity of direction			▨			The need for change was clearly explained. Desired outcome and project objectives were outlined.
	Performance measures		▨				Measurements were available on desired outcomes, such as alignment with Corporate Strategy Priorities, risks/benefits and key stakeholders' engagements. Fewer measurements were available for milestones and deliverables or CSFs.
	Quality of IT/business plan		▨				Some details were given, including the estimation of cost/benefits, stakeholder and staff involvement in decision making, and issues of risks. Generally, the project covers required topics but in broad terms, lacking details.
Knowledge	IT/business manager communication		▨				Communication was in its beginning stage.
	Organisation emphasis on knowledge			▨			A transaction plan addressed broad issues, accountability, process documentation, communication plans, audit operation, the relationship with the Certifying Body, and corporate coordination and budget process. Needs for change were identified.
	Training	▨					Little attention was given to training.
Decision Making	IT investment and budgeting		▨				While decisions were made in a formal process, details funding was not allocated for this project and the total funding requested by BC was not mentioned. Other funding decisions made by directorates were identified.
	Prioritisation			▨			The project was undertaken based on strength of alignment to Corporate Strategic Priorities, alternative solutions, value-business outcomes and internal resource priorities
	Stakeholders			▨			Internal and external stakeholders were engaged, and their impacts and their requirements were addressed.
	Systematic decision		▨				Scope options, value/cost assessments and risk profiles were investigated with stakeholders

Enterprise Architecture	**Aligned technical/business solutions**						Technical and business solutions were integrated with directorate impacts by means of internal resources, and linked to the Corporate Risk Management Programme. The project is situated under the Business Management Systems within the Corporate Risk Management Programme.
	Application and technology						The project EA layers faced no difficulty integrating with other systems such as the Corporate Risk Management Programme.
	Risk assessment						Measurements taken included revision history; two risks, their likelihood of impact and ways to mitigate them; and a Benefits Register.
Public Value	**Benefits to organisation**						Benefits included one certifying body for the accreditation of BMS with the Council, a corporate reporting framework and standardised corporate deliverables developed by directorates themselves.
	Benefits to public						Planned benefits included efficiency gains and effectiveness operation, which will in turn improve customer services of the Council.
	Economic/financial metrics						A table of cost estimation and VRP was produced.

TABLE 5-6: MATURITY OF SA ATTRIBUTES IN PROJECT 6

SAP	SA Attributes	Maturity					Rationale
		1	2	3	4	5	
Strategy	Clarity of direction						Objectives were clear and established, and included a plan to continue operational management of the OS Business Continuity Plan. Activities and resources were planned and provided with maximum notice of stakeholder involvement so as to minimise the resource impact particular to IT Operations resources.
	Performance measures						Performance measures included in/out scope, deliverables, constraints and dependencies, and project structure.
	Quality of IT/business plan						The business plan included a few details about project schedule, stakeholder analysis, impacts on other systems, change management, risks and issues, project funding details and external resources.
Knowledge	IT/business manager communication						Some communication occurred between them.
	Organisation emphasis on knowledge						There was an emphasis on sharing goals and responsibilities with all partners and stakeholders throughout the project.
	Training						Attention has not been given to training.
Decision Making	IT investment and budgeting						Resource estimate decisions were made in accordance with the directorate group and with Business Continuity Plans 2005-2006. Budget/funding and value profiles were taken into consideration in determining project expedition.
	Prioritisation						Business continuity is necessary in the event of a service interruption.
	Stakeholders						Key project stakeholders were involved in decisions.
Enterprise Architecture	Aligned technical and business solutions						Technical and business solutions were integrated with other projects, forming part of the overall Corporate Risk Management Framework Implementation Program.
	Application and technology						Network enterprise applications include the Corporate Risk Register, various specific risk registers, the Risk Policy and Guidelines, the Business Continuity Plan and a Disaster Recovery Plan to ensure all of these various parts work in concert with each other and result in a consistent approach to the identification and mitigation of risk across the organisation.
	Risk assessment						A few measurements have been taken into consideration in BC and PMP, such as links to other projects, alignment impacts, cost/benefits estimates, deliverables and project risk management

Public Value	Benefits to organisation						Business continuity offered benefit to both the organisation and to society.
	Benefits to public		☒				Benefits included improving public services through continuity of the Council's business even in the event of destruction (earthquake, bushfire, flood).
	Economic/ financial metrics		☒				A table of resource estimates was created to estimate a cost-benefit analysis.

TABLE 5.7: MATURITY OF SA ATTRIBUTES IN PROJECT 7

SAP	SA Attributes	Maturity					Rationale
		1	2	3	4	5	
Strategy	Clarity of direction				☒		Objectives were explained and distributed, and included the development of Council Planning Scheme, Risk Smart Development Application, Risk Assessor, Legal requirements and Customer Access to Approved Building Application.
	Performance measures				☒		Performance measures were applied to each above objective; identified activities were required to secure and sustain process changes in the Council. Self-evaluation check lists were carried out; deliverables, scope, and project closure reports were implemented to ensure project performance.
	Quality of IT/business plan			☒			Details about functionality requirements of the project were explained: cost estimation, risk descriptions and consequence scale, performance measures via three activities' streams, alignment to ICT, Corporate priorities and in/out scope of the project. Links and interdependences, however, were not identified.
Knowledge	IT/business manager communication						Little communication occurred via the Project Steering Committee and Project Management Group.
	Organisation emphasis on knowledge			☒			The main focus of the project was communicated, along with responsibilities and stakeholder engagement. Intranet communications and RRIF PD Online communication plans have ensured that the project, solutions and its benefits have been communicated to other directorates, to the development industry, to the Council and to the public. An extensive OCM and communications plan for the project was developed and implemented. A memo to the CEO proposed that the ownership of the solution is a higher strategic level of governance, since multiple directorates use the solution.
	Training			☒			The project demonstrated ongoing support solutions for the PD Online; maintenance and support of the RRFI PD Online training environment; significant training for PE and T and other directorates; operational support training in BSU; and Master Plan Editor training of SEPP planners.

SAP	SA Attributes	1	2	3	4	5	Rationale
Decision Making	IT investment and budgeting				▓		Decisions to allocate funds for the project were made, and breakdown issues and costs were finalised via the Infomaster's Planning Online Scheme. Decisions were made during the course of the project because it was felt that e-Plan tool was inferior to what the Council already offers through its online Planning Scheme.
	Prioritisation		▓				Decisions were prioritised based on strategic alignment, on corporate strategic priorities and ICT strategy and on cost and value profiles.
	Stakeholders			▓			Stakeholders were engaged, but internal stakeholders and interested parties were not.
Enterprise Architecture	Aligned technical/business solutions				▓		Network capacity of IT/business solutions integrated with the current architecture of PD Online is sufficiently robust to cope with the increase load of PD Online. Applications for legal requirements and customer access were built.
	Application and technology			▓			A diagram of a major integrated functionality plan was created, along with supporting application.
	Risk assessment			▓			The Project Risk Register outlined and considered risks. Out of 64 risks, two risks remained open.
Public Value	Benefits to organisation			▓			Benefits included: improving business process, alleviating business risk, bankable products, avoided costs and demonstrating efficiency gains.
	Benefits to public				▓		PD Online offered functionality to the public through the Council website. Benefits included improved customer service, free public access to Planning Scheme and building and development section within the Council.
	Economic/ financial metrics		▓				Traditional economic/financial metrics were less present.

TABLE 5-8: MATURITY OF SA ATTRIBUTES IN PROJECT 8

SAP	SA Attributes	Maturity					Rationale
		1	2	3	4	5	
Strategy	Clarity of direction				▓		The primary objective is to roll out new telecom services in three subprojects: voice services, data and internet services and mobile communications services. Principle objectives of EOI and organisational goals were identified and communicated in BC and PMP.
	Performance measures				▓		Evaluation criteria were used to suit the solution most advantageous to the Council. Requirement rankings in the tender process were implemented. Performance measures, delivery strategy, general issues, deliverables and expected benefits were also addressed.
	Quality of IT/business plan				▓		Technical specification was identified based on business requirement provided by representatives of each of the directorates. Milestones, constraints and dependencies, in/out scope, resource estimates, stakeholders analysis and related initiatives were explained in detail.

Knowledge	IT/business manager communication						Due to the nature of the project, communication was common.
	Organisation emphasis on knowledge						A project members list and document history were distributed; stakeholders and members of another project team were informed of planning stages and the implementation process. Communications management was ongoing throughout the project. Performance reporting was described in table format (to whom, what, when, how and prepared by whom).
	Training						APET training was provided for Evaluation Group Members.
Decision Making	IT investment and budgeting						Decisions were well supported and aligned with financial benefits analysis. Budgetary impact, financial control, charges and payments, financials consumed to date and related project cost management were all taken into consideration.
	Prioritisation						Prioritisation based on cost benefits analysis The project had four strategic priority areas: customer service, leadership and governance, information and knowledge management and internal services. However, the project demonstrated conflicting priorities with the Burleigh Tower Relocation Project. -
	Stakeholders						Human Resource Management included all stakeholders in communications management, including sponsors, business owners and users, customers, team members, subject matter experts, vendors and interested parties (organisational planning, staff acquisition and team development). Performance reporting was described in table format (to whom, what, when, how and prepared by whom).
Enterprise Architecture	Aligned technical/ business solutions						High integration and high business technical integration specifications were present. the technical lead, voice engineer, data network engineer and business analyst were involved.
	Application and technology						Service specific requirements were left empty in PMP; however, technical assumptions and technical constraints were provided. The location of existing technology infrastructure was identified and telecommunications equipment were migrated from existing poles.
	Risk assessment						Risks were identified as high-, medium- or low-risk. Risk management consisted of a risk management plan, risk identification, qualitative risk analysis, quantitative risk analysis, risk response planning and risk monitoring and control. A health check evaluation was implemented.

Public Value							Rationale
	Benefits to organisational						Benefits focussed within the organisation and included the ability to pursue a bundled procurement process for telecommunication services via an EOI process; to deliver the most cost effective outcome for the Council; and to increase the operational effectiveness.
	Benefits to public						Benefits included various services to the community, including the GCW Nerang Call Centre, Customer Service Counters, water catchments, water purification and distribution, sewage collection and treatment, garbage collection, library services, health inspections, regulatory services, parks and gardens, etc.
	Economic/ financial metrics						Variation in time and budget was applied; information (such as roles, deliverables, commercial obligations) was conveyed and rationalised to stakeholders and team members.

TABLE 5-9: MATURITY OF SA ATTRIBUTES IN PROJECT 9

SAP	SA Attributes	Maturity					Rationale
		1	2	3	4	5	
Strategy	Clarity of direction						A BC document identified business need, showing, for example, a 172% increase in the number of visits, a 32% increase in loan transactions, 625 new memberships and a total of average 38,000 visits per month in 2003–2004. -The ICT model was reviewed and confirmed scalable for the library; costs were reduced by retiring ICT infrastructure components.
	Performance measures						Corporate Risks and issues, project alignment, option analysis, in/out scope, value profiles and key project stakeholders were identified.
	Quality of IT/business plan						Expected risks and benefits, resource estimates, options analysis of ICT infrastructure, in/out scope, key stakeholders, project alignment and alignment to other systems were explained in the BC, but more details were required to provide sufficient information. The project had no CSFs or training plan.
Knowledge	IT/business manager communication						Through directorates and stakeholder distribution, the business system owner, and steering committee and project team communicated with one another.
	Organisation emphasis on knowledge						Knowledge was shared throughout the project via stakeholders' meetings, steering committee and team members.
	Training						No training occurred, perhaps due to the nature of this project, where staff are familiar with and focussed on customer service.

	SA Attributes	1	2	3	4	5	Rationale
Decision Making	IT investment and budgeting			X			Decisions were taken based on alignment position, value profile and inputs from stakeholders and project members; decisions were taken by the steering group, project members and the structure of IT governance.
	Prioritisation			X			Prioritisation process was structured by criteria specified in IT investment and budget.
	Stakeholders		X				Around thirteen key stakeholders were involved in decision making, eight of whom required at least consultation in decision making.
Enterprise Architecture	Aligned technical/ business solutions			X			Flexible integrated design was applied to meet the needs of future service growth.
	Application and technology		X				An ICT infrastructure alternative was analysed; the team provided advantages/disadvantages of three options for preferred new ICT library infrastructure.
	Risk assessment		X				A table was created to estimate and describe risks and their mitigation.
Public Value	Benefits to organisation		X				Planned benefits included reduction in ongoing support via the utilisation of centralised support services, and flexible design which could evolve with the future growth of services.
	Benefits to public			X			Other outlined benefits included customer self-service facilities.
	Economic/ financial metrics	X					A table of project planning costs and a table of corporate risks and issues were used.

TABLE 5-10: MATURITY OF SA ATTRIBUTES IN PROJECT 10

SAP	SA Attributes	Maturity					Rationale
		1	2	3	4	5	
Strategy	Clarity of Direction				X		The need for the MIDAS system (to replace CABS system) was clearly explained in BC and PMP.
	Performance Measures			X			Performance measures included milestones and deliverables, taken baseline data, scheduled data, actual data, comments, contract preparation, requirement specification, classification report, risks register, critical success factors and project handover.
	Quality of IT/business plan		X				The IT/business plan was extensive and included scheduling, training, milestones and deliverables, activities and responsibilities, system integration, in/out scope, resource plan and stakeholders. However, these elements were addressed only broadly. PMP template was not properly structured and did not cover details. Project Quality Management was not in place, and Project Management Methodology was relatively new and immature.

Knowledge	IT/business manager communication					Communication was scheduled in a stakeholder plan and project manager reports to the project executive. The presence of many stakeholders caused conflicts among them.
	Organisation Emphasis on Knowledge					Extensive stakeholder engagement and responsibility sharing. A kick-off workshop was not attended; uncertainty about how the project was to be managed and monitored dominated the project team and members. No clear methodology was followed.
	Training					Train-the-trainer and risk management workshops were preferred but not highly attended. The use of Council systems, updated knowledge and skills and five backfill positions were provided by the Council. Skilled contract/negotiation personnel were not available.
Decision Making	IT Investment and Budgeting					The estimated budget changed dramatically from 1.6M in 2004 to $6M in 2006 and to $11.5M in 2007; the cost increased due to unplanned works. Decisions were taken by committees but were occasionally delayed and more costly.
	Prioritisation					The project was selected because the CABS systems reached its serviceable life. Priorities were to offer value to its customers and to operate effectively and efficiently.
	Stakeholders					Many stakeholders were identified. However, some members of group management were not stakeholders and conflicting interests and opinions arose over business requirements.
Enterprise Architecture	Aligned Technical/business Solutions					Technical and business solutions were integrated by the Council, Organisational Services and the GCW. Scope was integrated by business solutions, such as service offering, role and responsibilities, SLA and Key Performance Indicators and resources required.
	Application and technology					Doubts arose about MIDAS's capability to interface with other corporate systems, which would have an impact on infrastructure services and application supports. The preferred solution exceeded reasonable tolerance levels.
	Risk assessment					The Risk Management Plan, Risk Control Register and the Risk/Issue/Change Register were used to identify risks in management and at any point of time. Technical issues and design phase faced many options and requirements to flesh out risks. The majority of risks pertained to technology, people and management.

Public Value	SA Attributes						Rationale
	Benefits to Organisation			▓			A few objectives were outlined: Cisco Secure Licenses, Telstra CDMA Network purchased, Alpha West survey and Cisco Radius Server, a few accounts activated in CDMA network, etc. Outlined benefits to the organisation were given.
	Benefits to Public		▓				Benefits to public services were introduced in IT plan.
	Economic/Financial Metrics		▓				BRP was used to measure bankable benefits, costs, and efficiency gains. Other metrics did not exist.

TABLE 5-11: MATURITY OF SA ATTRIBUTES IN PROJECT 11

SAP	SA Attributes	Maturity					Rationale
		1	2	3	4	5	
Strategy	Clarity of direction			▓			Objectives were clear and integrated with other sub-projects of the Information Management Program. The main objective was to provide a single reference source for information to access rights, personnel and stakeholders, as well as to locate all Council information. Specifically, the objective was to provide a plan for significant purchase for the procurement for the ECM. Procurement objectives were all identified.
	Performance measures			▓			A number of measurements were been taken: risk quantification, risks and issues, deliverables of key milestones, contract/performance management, start and finishing dates, CSFs and risk assessment.
	Quality of IT/business plan			▓			The project was well documented and had all three main documents, such as BC, PMP, and PIR. Details of options analysis, benefits/costs, scope, assumptions, policies, standards, guidelines, deliverables and key themes were addressed.
Knowledge	IT/business manager communication		▓				Little communication occurred.
	Organisation emphasis on knowledge		▓				The need for change was communicated, and the aim of creating systematic approach to managing all information systems was shared among organisation employees. The project moved from knowledge-blocked to knowledge-centred
	Training			▓			Change management included training and communications. Change management and the communications team interacted effectively to ensure that internal and external communication was meaningful, sensitive and timely. 120 staff were selected across the Council for Test Deployment of the ECM.

Decision Making	IT investment and budgeting						Most formal decisions were taken based on cost analysis: directorate personal cost, project personal cost, ongoing support costs, risk contingency, equipment and infrastructure cost, cumulative cost/benefits, significant purchase, RFT/RFI and estimate scope of purchase.
	Prioritisation						Three options were available. The third option was chosen because it has greater business and operational benefits while managing cost and risk in development of PMP for each of the four sub-projects. The project was aligned to Strategic Priority and corporate priorities. Procurement options were selected.
	Stakeholders						Stakeholder analysiswas carried out, and stakeholders, including suppliers and industry representatives, were engaged. Project team and project management members were also included.
Enterprise Architecture	Aligned technical/business solutions						The IM Program operationalised the ICT Strategy 2005/2009 and aligned to the Corporate Plan 2005-2009.
	Application and technology						The project established a platform of tightly integrated enterprise-capable applications and functionality which flexibly supports Council's changing business; can related all e-information into a single repository that can migrate the Council to a consolidated solution that suits other systems; increased the capability maturity level of information management; and created a combined internet/intranet/presence that will begin enabling personnel and stakeholders to access information from anywhere (remote access).
	Risk assessment						Three options to mitigate risk issues (cost/benefit) were addressed. A corporate framework was developed for economic and risk-free investment strategy. The need for change to protect information management was addressed.
Public Value	Benefits to organisation						The organisation's desired outcomes were outlined in BC (avoided cost, productivity, bankable benefits, improved quality and reduced risk). The benefits generated from IM are a result of the integration of four projects with critical dependencies, preferable to projects in isolation. However, little baseline data is shown in the benefit table. Other benefits were identified, including storage requirements, increased system scalability, increased document management efficiency and file security and auditing.
	Benefits to public						Benefits included improving customer service and online service availability.
	Economic/financial metrics						Program breakeven metrics and VRP were used.

TABLE 5-12: MATURITY OF SA ATTRIBUTES IN PROJECT 12

SAP	SA Attributes	Maturity					Rationale
		1	2	3	4	5	
Strategy	Clarity of direction						Objectives were clearly defined and aligned to the main objectives of the whole IM Program, which were to create a single and complete record of all information under Council's control; to make finding relevant and accurate information easier; -to build a single-stop corporate framework for information from anywhere and by all employees; and in general to offer shared knowledge and transparency for all.
	Performance measures						CSFs, risks and issues, deliverables and desired outcomes, milestones , and Program Kick Off were addressed BRP was also created to address benefits in terms of effectiveness, efficiency and money value.
	Quality of IT/business plan						The project has BC, PMP, PIR and VRP; proposed scope/ out scope, Audit Data Maintenance, transition approach and technical asset analysis were also conducted. Issues covered included deliverables, milestones, tangible and intangible benefits, cost/benefit estimation options, performance measures, stakeholders, risks, links and interdependencies, corporate strategic priorities and directorate impacts, but the issues were not all covered in enough detail.
Knowledge	IT/business manager communication						Change Management Communications aided business/ IT communication, which allowed for effective interaction among team members and stakeholders across the organisation.
	Organisation emphasis on knowledge						The Information Audit is part of the IM Program, which aims to focus on shared knowledge. Information assets are preserved in a usable form for the benefit of present and future generations. Formal and informal communication was identified to manage all information assets in order to allow all authorised generations to gain access. A Self-Assessment Checklist was used to shift from the project from knowledge-blocked to knowledge–centred. Responsibilities were shared by stakeholders and the entire project team; CG structures were informed.
	Training						Kick-off training was given to all staff and stakeholders. Training focused on shared information in the process in order to break down resistance.
Decision Making	IT investment and budgeting						IT action was taken based on the preferred solution chosen from among three options. Breakeven, internal resources, cost/benefits analysis were used to improve decisions. IT investment decisions were highly aligned with the future vision of organisation and its priorities.
	Prioritisation						Economically feasible and corporate strategic priorities.
	Stakeholders						Most staff and stakeholders were involved in shared responsibilities, decisions, directorate impact and knowledge. Issues of project personnel and distribution of key stakeholders were addressed.

SAP	SA Attributes						Rationale
Enterprise Architecture	Aligned technical/ business solutions						Integration was in its initial stage; specifically, business/ IT integrated with three other projects to form a holistic approach adhering to organisation vision.
	Application and technology						EA System Architect Implementation Meta Model Requirement Definition and EA Tool were created and made easier with the process of implementation. Business Analysis Services migrated new process patterns into the EA Tool. -EA conceptual framework identified the key information types and represented the functional model, such as application, classification, technology, technical services and business architecture.
	Risk assessment						Risks addressed included benefits/business outcome, performance measurement, risk and issues, cost estimation and cost/benefits analysis A project charter was not created (a PMO requirement). The confidentiality of data sensitivity measurement was still low.
Public Value	Benefits to organisation						Benefits to the organisation are many; the desired outcomes and the scope of the project were outlined in BC and PMP.
	Benefits to public						The information assets will be effectively managed and be available to the public, which will increase online customer services capabilities.
	Economic/ financial metrics						Net Present Value Calculations and BRP template were used.

TABLE 5.13: MATURITY OF SA ATTRIBUTES IN PROJECT 13

SAP	SA Attributes	Maturity					Rationale
		1	2	3	4	5	
Strategy	Clarity of direction						Objectives are communicated and aligned with those of the IM Program. Objectives were to create a single source of information that can be accessed under secure conditions, by anyone from anywhere.
	PERFORMANCE MEASURES						Performance measures included CSFs, risks/issues, deliverables/milestones, benefits/cost measurement.
	Quality of IT/business plan						The project lacked a PIR document. Issues such as risk analysis, cost/benefit analysis, milestones, business analysis and time line deliverables were addressed in detail. Three alternative options were considered, in scope and out scope.-

Category	Criterion						Description
Knowledge	IT/business manager communication						Communication was included in program management; the IM program governance had a management team which includes the CIO, ELT, steering committee, and IT/Business Reference Group.
	Organisation emphasis on knowledge						The aim of the project was clearly communicated to all stakeholders and staff. Various team members were engaged, including ELT, CIO, IT/Business managers, vendors and all stakeholders. The goal was to manage, spread and share information to and with all staff and citizens.
	Training						A kick-off workshop took place for all staff and stakeholders. Communication Strategy increased awareness and understanding of the program and ensured employees and stakeholders knew where to find further information.
Decision Making	IT investment and budgeting						Key decisions were taken based on cost/benefit estimation options (breakeven), ICT ten-year budget estimates, a value profile, etc.
	Prioritisation						Decisions were based on corporate strategic priorities, the corporate plan 2005-2009, a Business Plan from the office of CIO, ICT strategy 2005-2009 and risk/benefit analysis.
	Stakeholders						Since human resources will be much impacted by the IM program, engaging stakeholders in OCM was crucial. Directorate impacts also encouraged the participation of stakeholders, staff and other member in the decision-making and implementation process.
Enterprise Architecture	Aligned technical/business solutions						Technical/business solutions complied with the EA principles. iSPOT, the web platform and the ERP and property systems, accomplished the integration of solutions and components.
	Application and technology						A new architect was introduced within BA and OCIO for tracking and reporting on policy uptake eLearning capability. Project components improved the capability of the organisation.
	Risk assessment						CSFs, risk estimation and value profile were used but significant variance in the estimates of risk mitigations were found by a number of contributors, deriving from a lack of firm basis for the estimate.
Public Value	Benefits to organisation						Major benefits included efficiencies in work processes and managing and disposing of information, effectiveness in service delivery and cost reduction in terms of time taken to find information or in storage costs.
	Benefits to public						The main objective was to create a single source of well managed information in order to improve customer services.
	Economic/financial metrics						Breakeven and business case benefits, tangible and intangible benefits were used as metrics.

TABLE 5-14: MATURITY OF SA ATTRIBUTES IN PROJECT 14

SAP	SA Attributes	Maturity					Rationale
		1	2	3	4	5	
Strategy	Clarity of direction						The objectives and the purposes of the Policy project were clearly identified: to ensure that information under the control of the Council is managed consistently, effectively and efficiently throughout its life cycle; to provide the right resources with the right competencies to meet business needs; to implement IM Policy Communication Plan and Information Privacy principles; and to protect both customer and Council information assets.
	Performance measures						IM Policy deliverables, CSFs, delivery of key milestones, risks/issues analysis, project deliverables and consequence scale of deliverables in terms of probable, possible, and improbable were all identified. Management of risk, risk assessment, internal resource estimations occurred.
	Quality of IT/business plan						Most issues were addressed in detail, including cost estimation options analysis, in scope and out scope, milestones delivery, definition of terms, roles and responsibility of stakeholders and team management, compliance with Queensland Information Standards IS31, record-keeping, contractual arrangement with vendors and implementation steps, as well as constraints and dependencies. Plan communicated legislation, project schedules and related standards in terms of capability of technology infrastructure.
Knowledge	IT/business manager communication						Discussion took place with staff from all directorates in the Council, including members of the IM group and the TRACKS project leader.
	Organisation emphasis on knowledge						Sharing knowledge is the focus and strategic priority of the IM Policy Project, as a sub-project within the IM Program. -Communication, Change Management and Benefits Realisation Plan were present. -IM focussed on everyone's accountability for managing information, on a consistent disciplined approach across the Council and on flexibility and cooperation in terms of the Policy Project's involvement in the IM program. Defining the scope and sharing responsibility was emphasised.
	Training						An Education Plan, emphasis on knowledge awareness for all, and kick-off training increased commitment to the organisation.

Decision Making	**IT investment and budgeting**					The project was recommended based on a profile of its value to the organisation. Three options are outlined in the value profile and a Corporate Approach preference was chosen.
	Prioritisation					Option 2 was formally taken because it produced greater business and operational benefit while managing cost and risk. The project proceeded with development of PM Plans for each of the four project components. Other criteria used were equipment/infrastructure, cost estimates, strategic alignment and corporate capability.
	Stakeholders					Since the impact appears across organisations, all directorates and stakeholders (via stakeholder analysis) were engaged in the key decisions and implementation process. Responsibilities were shared via a Policy Instrument document
Enterprise Architecture	**Aligned technical/business solutions**					The project was aligned with Strategic Priority 16 and with Business Plan (Audit, Continuity Integrated and Policy) and integrated with other systems such as ICT Strategy, Business Plan and Corporate Plan 2005-2009.
	Application and technology					The project was aligned with ICT Strategy, Corporate and the Business Plan make easy integration with other systems across the organisation. Quality of integrity of information Assets was ensured.
	Risk assessment					Risks and issues were addressed in table format (for example, lack of senior management support, conflicting priorities, focus on technology rather than business requirements). Other assessments were in the form of cost/benefits estimation analysis, in scope and out of scope and key delivery of milestones.
Public Value	**Benefits to organisation**					The Benefits Register and Benefits Realisation Plan were used to define benefits to the organisation. The need and the purpose of the project was used to identify benefits.
	Benefits to public					Planned public benefits included improved customer service, improved business process and efficiency gains, improved accountability/transparency, and compliance with Queensland Government Standards.
	Economic/ financial metrics					Tangible benefits measurement, benefits/business outcomes and value profile via breakeven were applied in project plan.

APPENDIX J: ATTRIBUTES DISTRIBUTION ACROSS PROJECTS

FIGURE 6-2: CLARITY OF DIRECTION: ATTRIBUTE VALUE ACROSS PROJECTS

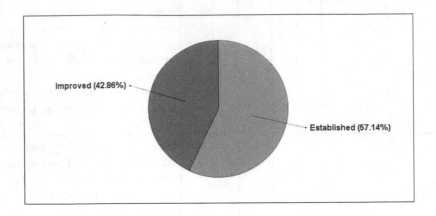

FIGURE 6-3: PERFORMANCE MEASURES: ATTRIBUTE VALUE ACROSS PROJECTS

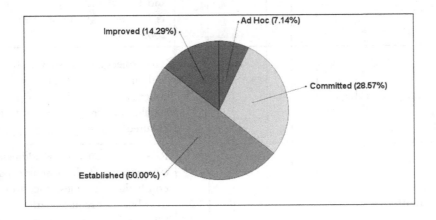

FIGURE 6-4: QUALITY OF IT/BUSINESS PLAN: ATTRIBUTE VALUE ACROSS PROJECTS

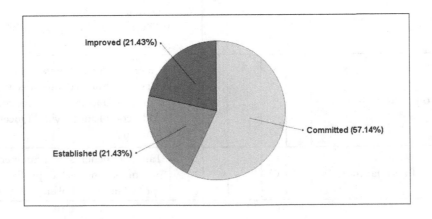

FIGURE 6-5 IT/BUSINESS MANAGERS' PARTICIPATION: ATTRIBUTE VALUE ACROSS PROJECTS

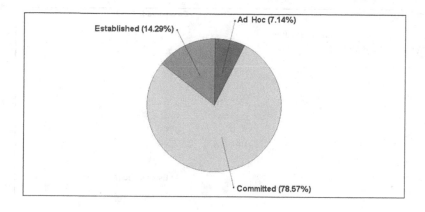

FIGURE 6-6: ORGANISATION EMPHASIS ON KNOWLEDGE: ATTRIBUTE VALUE ACROSS PROJECTS

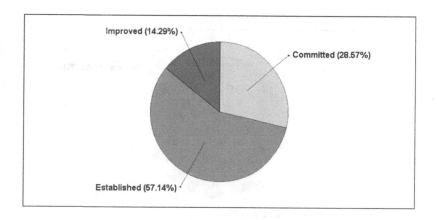

FIGURE 6-7: TRAINING: ATTRIBUTE VALUE ACROSS PROJECTS

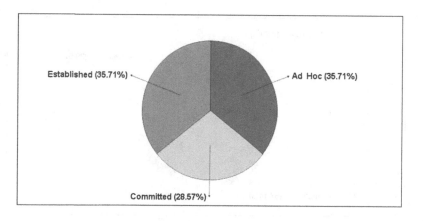

FIGURE 6-8: IT INVESTMENTS AND BUDGET: ATTRIBUTE VALUE ACROSS PROJECTS

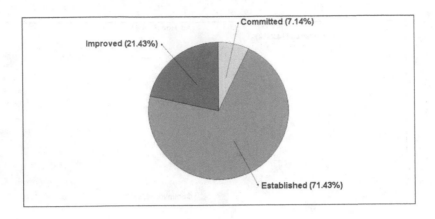

FIGURE 6-9: PRIORITISATION: ATTRIBUTE VALUE ACROSS PROJECTS

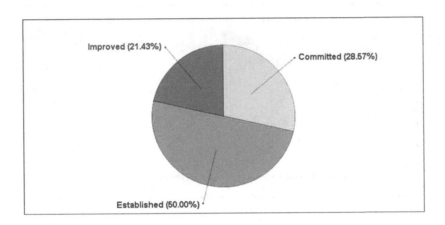

FIGURE 6-10: STAKEHOLDERS: ATTRIBUTE VALUE ACROSS PROJECTS

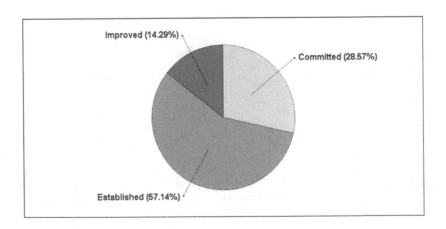

FIGURE 6-11: ALIGNED TECHNICAL/BUSINESS SOLUTIONS: ATTRIBUTE VALUE ACROSS PROJECTS

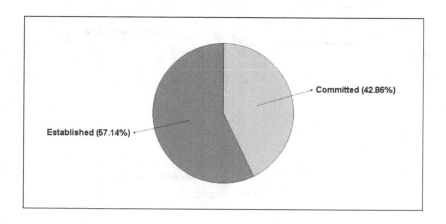

FIGURE 6-12: APPLICATION AND TECHNOLOGY: ATTRIBUTE VALUE ACROSS PROJECTS

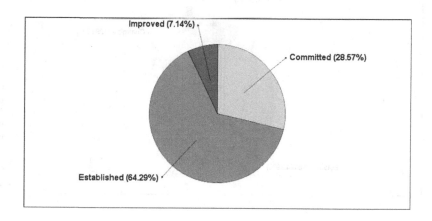

FIGURE 6-13: RISK ASSESSMENT: ATTRIBUTE VALUE ACROSS PROJECTS

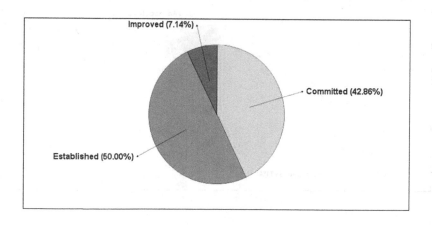

FIGURE 6-14: BENEFITS TO ORGANISATION: ATTRIBUTE VALUE ACROSS PROJECTS

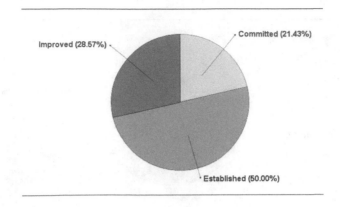

FIGURE 6-15: BENEFITS TO PUBLIC: ATTRIBUTE VALUE ACROSS PROJECTS

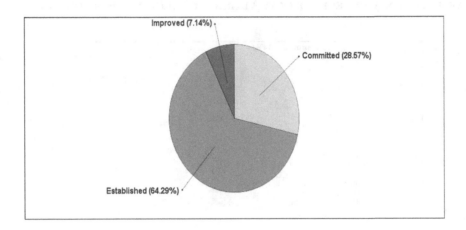

FIGURE 6-16: ECONOMIC/FINANCIAL METRICS: ATTRIBUTE VALUE ACROSS PROJECTS

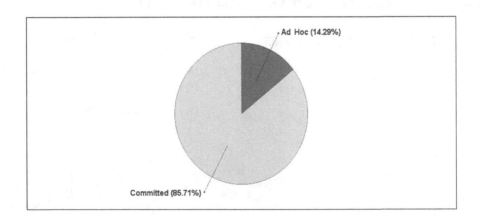

Appendix K: Attributes maturity and success rate relation

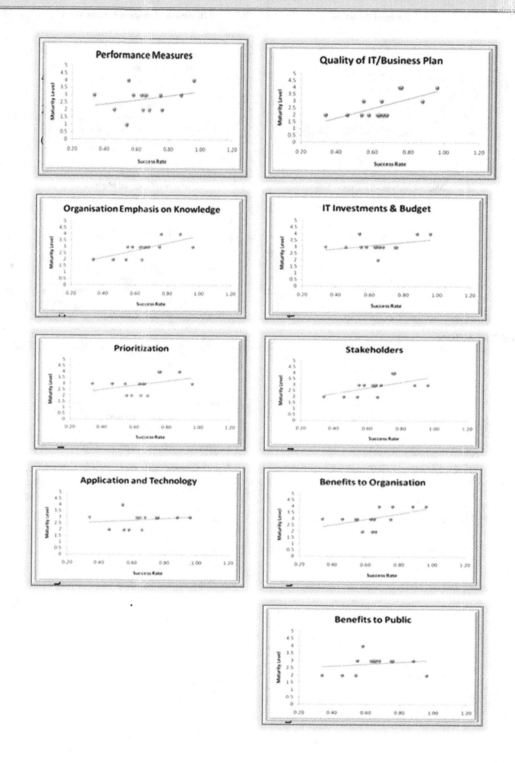

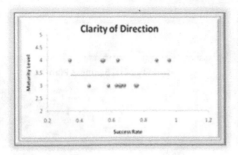

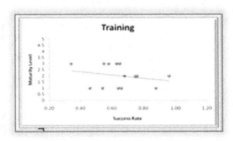

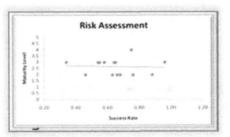

APPENDIX L: TREE NODES USING NVIVO

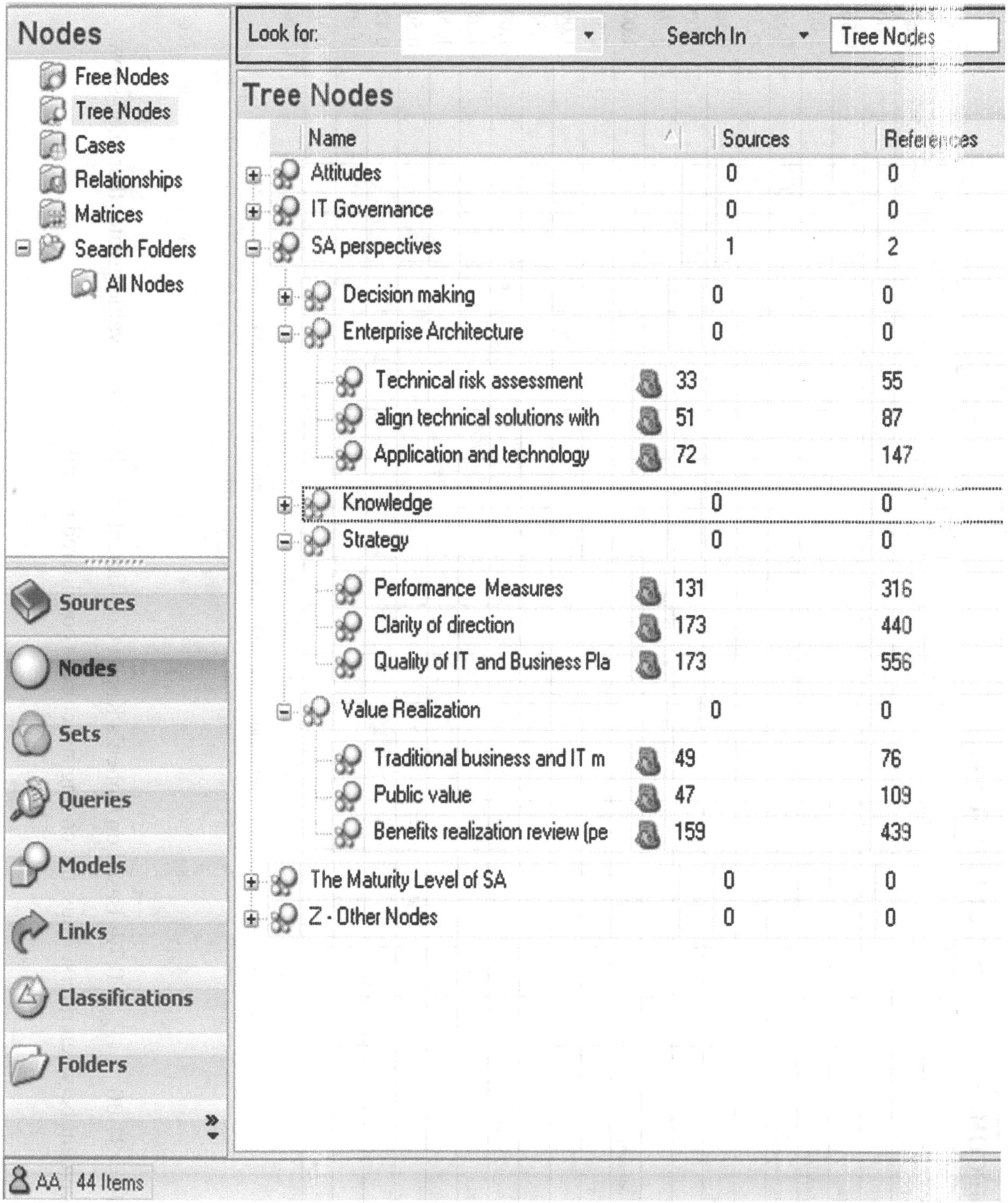
Nodes
Free Nodes
Tree Nodes
Cases
Relationships
Matrices
Search Folders
All Nodes
Look for:
Search In
Tree Nodes
Tree Nodes
Name
Sources
References
Attitudes 0 0
IT Governance 0 0
SA perspectives 1 2
Decision making 0 0
Enterprise Architecture 0 0
Technical risk assessment 33 55
align technical solutions with 51 87
Application and technology 72 147
Knowledge 0 0
Strategy 0 0
Performance Measures 131 316
Clarity of direction 173 440
Quality of IT and Business Pla 173 556
Value Realization 0 0
Traditional business and IT m 49 76
Public value 47 109
Benefits realization review (pe 159 439
The Maturity Level of SA 0 0
Z - Other Nodes 0 0
Sources
Nodes
Sets
Queries
Models
Links
Classifications
Folders
AA
44 Items

APPENDIX M: Cohen's Kappa coefficient was used to measure the reliability of the coding structure normally known as nodes in NVivo. Four ratters were asked to categorize 20 quotes into the most appropriate nodes. The four raters included a Professor, an Assistant Professor, a PhD candidate and me. The results are given in the table below.

APPENDIX M: COHEN KAPPA'S COEFFICIENT

	A	B	C	D	E	F	G	H	I	J	K	L	M	N	O	P	Q	R	S	T	
Performance Measures:	1										3										0.42
Clarity of Direction:													4								1.00
Quality of IT/Business Plan:			1												3						0.50
IT/Business Managers' Participation:					4					0											1.00
Organization Emphasis on Knowledge:																	3	1			0.50
Training	3															1					-0.25
IT Investments & Budget:												3							1		0.50
Prioritization:				4													0				1.00
Stakeholders:														1						3	0.50
Align Technical/Business Solutions		4																			1.00
Application and Technology:						3														1	0.50
Risk Assessment:															0				4		1.00
Benefits to Organization:							1	3													0.50
Benefits to Public:																0		4			1.00
Economic/Financial Metrics		1												4							1.08
Strategy:							3					0				4					1.00
Knowledge:								1													0.50
Decision-Making:	1		3																		0.42
Enterprise Architecture:										3									1		0.50
Public Value/Value Realization Plan:								3					1								0.50
	5	5	4	4	4	3	4	3	4	3	3	3	5	5	3	5	3	5	6	4	81.00
	0.0	0.0	0.0	0.0	0.0	0.0	0.0	0.0	0.0	0.0	0.0	0.0	0.0	0.0	0.0	0.0	0.0	0.0	0.0	0.05	0.05
	6	6	5	5	5	4	5	4	5	4	4	4	6	6	4	6	4	6	7	4	0.05
	0	0	0	0	0	0	0	0	0	0	0	0	0	0	0	0	0	0	1	0	

0.73

0.72

$$\kappa = \frac{\bar{P}-\bar{P_e}}{1-\bar{P_e}}$$; $\bar{P} = 81.00$; $\bar{P_e} = 0.72$ Based on Cohen's Kappa Coefficient result (0.72), a value greater than 0.60 shows substantial agreement between raters. Thus, the coding structure is considered reliable.

214

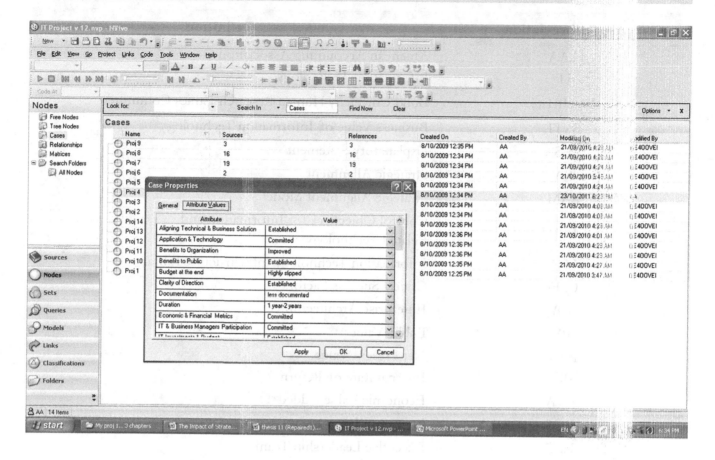

10. ABBREVIATIONS

Abbreviation	Definition
AAA	American Accounting Association
IT	Information Technology
BUHREC	Bond University Human Research Ethics Committee
BVIT	Business Value of Information Technology
ES	Explanatory Statement
SA	Strategic Alignment
SAM	Strategic Alignment Model
COBIT	Control Objectives for Information and Related Technology
ITIL	Information Technology Infrastructure Library
CSFs	Critical Success Factors
EA	Enterprise Architecture
PV	Public Value
NPV	Net Present Value
IRR	Internal Rate of Return
EVA	Economic Value Added
ROI	Return on Investment
ELT	Executive Leadership Team
CGC	Corporate Governance Committee
PMC	Portfolio Management Committee
OCIO	Office of Chief Information Officer
CIO	Chief Information Officer
PMO	Portfolio Management Office
BSRG	Business Solution Reference Group
TRG	Technical Reference Group